Coin Banks

by Banthrico™

James L. Redwine

Schiffer® Publishing Ltd

880 Lower Valley Road, Atglen, PA 19310 USA

Library of Congress Cataloging-in-Publication Data

Redwine, James L.
Coin banks by Banthrico / James L. Redwine.
p. cm.
ISBN 0-7643-1280-4
1. Banthrico, Inc.--Catalogs. 2. Coin banks--Collectors and collecting-United States--Catalogs. I. Title.
NK4698. R43 2001
688.7'2--dc21
00-012845

Designed by Bonnie M. Hensley
Type set in Impress BT/Lydian BT

ISBN: 0-7643-1280-4
Printed in China
1 2 3 4

Published by Schiffer Publishing Ltd.
4880 Lower Valley Road
Atglen, PA 19310
Phone: (610) 593-1777; Fax: (610) 593-2002
E-mail: Schifferbk@aol.com
Please visit our web site catalog at **www.schifferbooks.com**
We are always looking for people to write books on new and related subjects. If you have an idea for a book, please contact us at the above address.

This book may be purchased from the publisher.
Include $3.95 for shipping. Please try your bookstore first.
You may write for a free catalog.

In Europe, Schiffer books are distributed by
Bushwood Books
6 Marksbury Ave. Kew Gardens
Surrey TW9 4JF England
Phone: 44 (0)20 8392-8585; Fax: 44 (0)20 8392-9876
E-mail: Bushwd@aol.com
Free postage in the UK. Europe: air mail at cost.
Please try your bookstore first.

This book is lovingly dedicated to:

Delma Lewis Redwine, my mom, who was an antiques collector and got me started collecting toys as a boy,
Paullette Lohmann Redwine, my wife, who supported me through "the book adventure,"
Jim Darr, a close bank friend, who helped me at the beginning of the book with his fabulous Banthrico collection and who recently passed away,
Don Duer, bank collector, author and friend, who provided the "magical connection" between the author and the publisher,
Sandra Strohm, my typist, who did a magnificent job of turning my manuscript into the typed page,
the present and future generations of Banthrico collectors,
and to Jesus Christ, my personal Savior and Lord, who strengthened me daily in this literary challenge.

Purpose

This book's purpse is not intended to be the "last word" on coin banks by Banthrico. It is intended to be a fun adventure trip into Banthrico bank collecting. It is a journey through "virgin bank territory" because a majority of the Banthrico banks shown here have not been seen before by a majority of bank collectors, myself included. New banks will be discovered, more variations will come forward, and new history will be uncovered regarding Banthrico, Inc. Mistakes will be found and corrected. Let us together share the Banthrico story with future generations of collectors.

Please write or call me if you have new Banthrico information to share. We might need to update the book sometime in the future. Happy Bank Trails. James (Jim) L. Redwine, 6940 Lake Valley Drive, Florissant, MO 63033-5104, Tel. 314-741-6940

Acknowledgments

Encouragement for the development of this book was freely given by members of the Gateway Penny Bank Club (GPBC) in the metropolitan St. Louis area. It is a very active bank club with about 30 members, most of whom collect Banthrico banks. Thank you, one and all.

Fellow members of the Still Bank Collectors Club of America (SBCCA) and Souvenir Building Collector's Society (SBCS) also made extensive contributions to this book. SBCCA and SBCS members all over the United States , from Massachusetts to California, made contributions by providing lists, pictures, negatives, flyers, booklets, banks, and information. Without their help, much valuable information would be missing.

The SBCCA *Penny Bank Post* periodical, under the editorship of Bob Peirce, and the *SBCCA Newsletter*, edited by Mike Henry, have provided the network for disseminating and receiving Banthrico pictures and information. Their impact has been extremely helpful here.

The *Souvenir Building Collector*, Journal of the Souvenir Building Collectors Society (SBCS) guided by the editor, Dixie Trainer, also provided assistance by publishing a list of the building banks needed. Several members responded to the call and provided key building bank pictures and information. The author liked the organization and joined it.

Another group that has been extremely helpful is the "Banthrico Board" composed of twelve advanced Banthrico bank and building collectors selected by the author to rate the banks shown here. They are (in order of development): Jim Redwine, Missouri; Jim Darr, Missouri; Tony Carmi, Missouri; Andy Anderson, Illinois; Roy Jordan, Tennessee; Tom Hagan, Ohio; James Gass, Tennessee; Mike Sanko, Minnesota; Marty Gutierrez, Florida; Barry Hoffman, Massachusetts; Jamie Kessler, Texas; and Kent Burchsted, Texas. The Board members agreed to rate their own collections and any other Banthrico banks they wanted to rate. They used the SBCS/SBCCA Banthrico Building list and a list of their own banks. Some rated several hundred banks while others rated less than a hundred. A spreadsheet was developed for each chapter which included the banks and the twelve ratings.This approach allowed for input from veteran Banthrico collectors from eight geograhic areas. As was expected, the Board members did not always agree, but the author took the preponderance of ratings into consideration when assigning the final rating. Future ratings may change with the availability of more information. The author is very grateful to each member of the "Banthrico Board" for their efforts rating the banks.

The following people made significant contributions to the book. They spent hours with the author photographing their collections and provided extensive information, rarity and price lists, and input for the introduction. I salute you, fellow bank collectors and friends: Jim Darr, St. Louis, Mo.; Tony Carmi, Bridgeton, Mo.; Andy Anderson, Hinsdale, Il.; Roy Jordan, Cedar Grove, Tn.; and James Gass, Jackson, Tn.

There were so many who made very important contributions with one or several bank pictures and information. Thank you for your wonderful assistance (in chronological order of their assistance).
Larry Bates, Don Boswell, Bob Heberer, Fred/Edith Eads, Gordon Schroff, Bill Perrot, Will Saunders, Doug Harman, Michael Emrick, Guy Williams, Dean and Judy Rich, David Weingarten, Robin and Chas Wiebe, Peter Eisendrath, Margaret Majua, Tom Hagan, Curt Werner, Leonard Schenk, Don Erb, Julie Kidd, Gary Rosborough, Barry Hoffman, Jeane Morris, Joe Eichler, Mike Henry, Kent Burchsted, Edson Yetter, Tom Vick, Marty Gutierrez, Ed Wickham, Bob Peirce, Jamie Paul Kessler, Peggy Harder, Charles Steeber, Nancy Stroman, Michael Goldstein, Mike Boyle, Tom Stoddard, Bill Robison, Tom Kellogg, Larry and Jeanie Egelhoff, John Sherman, Don and Dorothy Gilbert, Mike Sanko, David Epstein, Bill Costello, Lance Oldham, Darryl Knutson, Karen and John Bennett, Gordon Steimle, Richard Young, Don and Betty Jo Heim, Dixie and Bill Trainer, Brian Cleary, Derek and Michelle Young, Carl White, Don Duer, N.M. Patton

There were others willing to help by sending their Banthrico collection lists. The rarity of the different banks could be gauged by their appearance on various lists.
Joan Lathouse, R. Hendrickson, John Marquand, Mario Franchi, Karen Kummer, Donald Speck, Susan Brown, Bob Feraco, A. Kenyon, Joseph Kopitz, Ann Mitchell, Bill Winter, Philip Hardacre, Mel Lipper, Monte Arnold, Nancy McGrath, Eupha Richie, Gen Luthultz, John Larson, Ned Bauer, Marilyn Feldman Walter Dreier, John Hill, Lynn Liebert, Steve Viola Al Goldstein, Richard Miller, Russ Allman

Anytime you start listing individuals who par ticipated in a project, there is always the chanc that someone will be omitted. If we failed to lis anyone who made a contribution or responded wit a list, we apologize, and please let the author kno so the mistake can be corrected.

Contents

Chapter 1
Introduction
The Banthrico Story, 1931-1985

Banthrico Inc. began in Chicago, Illinois, in 1931, on a foundation built by two companies before them. The name Ban-thri-co grew out of an older company, the Banker's Thrift Corporation, which Jerome "Jerry" Aronson and his brother-in law, Joseph Eisendrath, purchased with all its equipment in 1931 for $5,000. Banker's Thrift had recently acquired the Stronghart Company, in 1929, which had been in business since 1914. Both Stronghart and Banker's Thrift are well known as manufacturers of small pocket banks and small steel building banks.

The first president and co-founder of Banthrico Inc. was Jerry Aronson and the first secretary-treasurer and other co-founder was Joseph Eisendrath. They were partners in the bank manufacturing business for almost twenty-five years.

Banthrico continued the Stronghart and Banker's Thrift bank lines with the older names on the products for several years before the Banthrico name appeared on the banks. They used the old bank parts until they were depleted. Banthrico also used Stronghart's 1914 staring date as their company's beginning date.

Banthrico continued to grow and expand its line of promotional banks. They were making steel pocket and building banks and white metal banks in the 1930s. They purchased the assets of the National Products company in 1940, which included white metal model cars.

Just as Banthrico was going "full blast," World War II developed and Banthrico had to stop production of metal banks. Zinc, aluminum, brass, copper, and lead all were being used in the bank-making process but were needed for the war effort. Banthrico did its part during the war by making bomb fuses and radar parts through government contracts. During the war, Banthrico started making banks from plastic and coin folders from paper and cardboard in order to keep the "banking" part of the company going.

After the war, Banthrico once again made white metal banks by the thousands. They continued their autobank models with the National name and started putting their name on the bank models in 1949. You will find autobank models with National and/or Banthrico names on them during the change-over period, 1949 to 1951. There were busts of great Americans, mascots, animals, cars, trucks, boats, buildings, home appliances, food products, trophies, and a host of other miscellaneous categories. Banthrico had learned a lot about bank making from Stronghart, Banker's Thrift, and National Products and was poised for "full speed ahead" after the war. Just as the 1930s and 1940s were "preparation years" for Banthrico, the 1950s, 1960s, and 1970s were their "big production years."

When Jerry Aronson died in 1955, Joseph Eisendrath assumed leadership of Banthrico. Later, Jerry's son, Tom Aronson, became Vice-President of Sales and Joseph's son, Peter Eisendrath, became Executive Vice-President and Treasurer. Both Tom and Peter had grown up in the company with their dads and the Banthrico staff and they worked in many areas of the company. Peter's specialty was the process of making the banks and Tom's specialty was selling them.

In the 1970s, the company started making banks as bookends (the Lincoln Memorial is an excellent example) and adding music boxes to the banks (the Robot and Ford Taxi are good examples) to diversify their markets. Bookend and music box banks were sold directly to gift shops around the country. The music box mechanisms were made for them by Sanyo.

Banthrico increased its product lines again by purchasing Creative Castings of Addison, Illinois, in 1981. Creative Castings was in the business of casting metal parts for metal products manufacturers. Trophies, lamp parts, ornaments, figurines, and small machine parts were added to Banthrico's growing inventory of products.

When Joseph Eisendrath retired in 1982, Peter Eisendrath assumed the presidency of Banthrico. The 1980s were "recession years" for much of the American economy, including Banthrico.

Banthrico had been located in Chicago since its inception, first on W. Lake Street, then Des Plaines Street, west of the Chicago Loop and finally on Roosevelt Road on the West Side since 1970. The company employed around 75 employees, counting the office, sales, and plant staff, during their peak years. Banthrico used several company names during their history. First it was Banthrico, Inc., then Banthrico Industries, Inc., followed by Banthrico International Industries, Inc., and, lastly, back to Banthrico, Inc. They also had a division within the company in the 1980s entitled "Antiquo" which included the antique automobiles and the gift shop promotions.

In spite of their quality line of white metal banks and their diversification into other bank materials and products, it became evident that Banthrico was not going to survive. Banthrico was sold to Toystalgia, Incorporated, in 1985, which became Valley Casting Company in 1995.

Banthrico Banks Are Hot!

Interest in Banthrico banks has sky-rocketed in the last few years, along with their prices. Just a few years ago they were almost ignored by bank collectors. Now, they command keen competition. And they should — they are great works with intricate details that cover numerous subjects. You can expect Banthrico bank collecting to grow in the future.

This is a good field for beginning bank collectors to approach because the prices are still quite reasonable for the easier-to-find banks. And acquiring the harder-to-find Banthricos is a challenge to the veteran bank collector. In the future, there may be many Banthrico collectors because the availability, artistry, subject areas, investment potential, and growing price range of the banks is intriguing. White metal Banthricos have begun to take their rightful place in bank collecting circles, along with cast iron, tin, lead, and ceramic banks.

Banthrico ("The Coin Bank People" was their motto) has made a significant contribution to our country through their emphasis on thrift and saving money by adults and children alike. They marketed their toy banks to financial institutions and other organizations and individuals. The toy banks advertised the banking institutions and companies, often being given to their customers or sold for nominal fees ($3 to $5). Saving and thrift were encouraged by the financial community.

Banthrico made over 900 different banks. Most were made from white metal, but some early ones were steel, lead, celluloid, paper, cardboard, or plastic. Later banks also included vinyl plastisol and composition or paper mache. But, from the 1940s through 1985, the main Banthrico "bread and butter" banks were made of white metal.

The Banthrico banks were sold mainly to financial institutions, but were also marketed to furniture stores (Union National), industrial companies (Ralston Steel), appliance companies (Frigidaire), and automobile dealerships, to name a few. They eventually sold directly to gift shops and museums, such as Harrah's Automotive Museum in Nevada. Close to 50% of their total white metal bank sales came from their automobile molds. 95% of their output was sold in the United States and 5% distributed among Latin America and Europe, with Latin America getting the larger portion. The Banthrico records related to sales and bank production are incomplete today, which seems to be true of many companies that have closed their doors.

Since they are primarily made of zinc and aluminum and are prone to oxidation, many Banthrico white metal banks have not survived. Also, they are easily bent. This accounts for the scarcity of some issues. The banks need to be kept clean and dry; do not store them in a damp room.

Banthrico banks were mainly ordered in lots of 50, 150, 250, 500, 1000, and up. The majority of the orders were for generic banks that were ready to have an advertisement place on them, usually by the silk screen process of printing. Customers sometimes had Banthrico design and make a special order of a specific building, tower, monument, mascot, etc. in lot quantities that started around 500 or 1,000 banks. Over 5,000 banks were made for the First National Bank of Chicago and 5,000 John Deere tractor signs were made for the John Deere Credit Union in Iowa. Most special orders were for 1,000 to 3,000 banks. The ordering organizations owned the mold to their designs.

The majority of banks designed as busts, animals, and automobiles were generic and were made in the thousands. Good examples of these are the Lincoln busts, different pigs, and the Ford cars. The Thomas Dewey bust is scarce now and it may have been made as a small special order, or the mold may have been destroyed after Dewey lost the U. S. presidential election in 1948. The bank of a two-story dream house with a car in the garage is plentiful; perhaps the house and car was promised to voters by a hopeful presidential candidate..

Traps

Occasionally, banks will be found that are exactly alike, but one trap will indicate "Banthrico" and the other trap will indicate "A.C. Rehberger" or some other company. Some companies may have exchanged molds with each other, copied another company mold without their permission, or an enterprising individual has changed the trap from one bank to another because many traps are interchangeable. Many Banthrico traps are also interchangeable within their own products and they advertised their banks with traps having plain plates or key-locked plates. The Will Rogers bust bank is an example having both Banthrico and A.C. Rehberger traps.

Among the more than 900 different banks, there are 43 different traps in all types of materials including steel, celluloid, tin, lead, plastic, white metal, chalk, paper/cardboard, composition, and vinyl plastisol.

Blanks

Many Banthrico banks were sold as "blanks" with no advertisement on them. These were stock generic banks which Banthrico already had made up. They especially sold "blanks" in the 1980s to gift shops, museums, auto dealers and many had music boxes in them. College mascot banks were also sold directly to college bookstores without any advertisement.

Parts

Banthrico also contracted out the manufacturer of some banks and bank parts. Niagara Plastics made a lot of their plastic plugs, Master Lock Co. may have done some of their keys, manufacturers in Japan did several of their composition banks (Banthrico calls them "paper-mache"), such as Bucky Badger, and several companies did their vinyl plastisol banks, such as Paragon, who made Mickey Mouse.

Materials

When Banthrico started to cast metal banks, they used a heavy lead content in their metal mix. The building banks of the Continental Assurance Company of Chicago and the First Federal Savings and Loan Association of Detroit are heavy lead banks with excellent details. Their traps are both early, which indicates they were some of the first building banks Banthrico made. The white metal banks that followed are 95% zinc and 5% aluminum, with the slight addition of copper, brass and lead. (Some might argue that they contain silver and gold by their prices.) Banthrico calls them "slush molds" and some have called them "pot metal," but most collectors refer to them as "white metal" (WM) banks, as they are referred to here.

Some prices of Banthrico generic banks that were sold in lots:

	March 1972	January 1980
Antique autos	160 at $2.60 ea.	160 at $4.34 ea.
Bust bank	160 at $2.44 ea.	
Covered wagon	160 at $2.55 ea.	160 at $4.54 ea.
Stave barrel	160 at $2.04 ea.	160 at $3.00 ea.
American Home	160 at $1.12 ea.	
Clown, mechanical	150 at $1.60 ea.	
Coin folders	1250 at $54.24 /k	
Robot	160 at $4.49 each	
Little Lu Lu	160 at $2.50 each	
Sleeping Santa (vinyl)	160 at $2.50 each	
Drive-In Teller	160 at $1.99 each	
Variety Special		40stock banks @ $245 (about $6per bank)

Any assortment desired:

Bust Banks (Men of History)
Antique Autos
Americana series
College Mascots
Miscellaneous

Banks could be purchased in lots of 160, 320, 520, 1,000 and more. As the quantities increased, the prices were reduced a few cents per bank.

Numerous Banthricos in the 1960s, 1970s and 1980s could be purchased from banks for a few dollars with their names imprinted on them. These toy banks were profit centers and advertised the businesses at the same time. The bank's price was often reduced for opening special accounts and certificates of deposit.

Other Products

Banthrico also made a wide selection of cast bronze and aluminum signs, name plates, desk sets, paperweights, etc. for financial institutions. They were cast in one solid piece with raised letters and borders. They also made Federal Deposit Insurance Company (FDIC) and Federal Savings and Loan Insurance Company (FSLIC) emblems for interior and exterior financial institution displays.

In the 1960s, the following prices of specialty products were:
Bank Sign: 15"x20," $ 99.50 and 32"x36," $355.
FDIC/FSLIC emblems: 8," $ 39.50 and 14," $ 92.

These specialty items were expensive for the 1960s. They are closely related to bank collecting and of particular interest to some bank, building, and financial institution collectors.

The Rating System

The alphabetical system (first time system has been given a title) has gradually developed in bank collecting circles to rate banks and assist collectors and dealers in knowing what is common and what is rare. Rating banks with the alphabetical method, from A (most common) through F (most rare), has been in existence for over sixty years. In a now-out-of-print book entitled *Old Mechanical Banks*, copyright 1940, author Ina Hayward Bellows uses the alphabetical method of rating the mechanical banks. (The A banks were priced at $3 and the F banks were priced at $75.) Hubert "Bert" Whiting, in his out-of-print book, published in 1968, *Old Iron Still Banks*, uses the A through E rating method, dropping the F rating. It is considered a classic among still bank collectors and quite a collector's item. Andy and Susan Moore, in *The Penny Bank Book*, used the alphabetical method in rating still banks. Don Duer, in *Penny Banks Around the World*, used the alphabetical method and so does Tom Stoddard in his (1997) *Ceramic Coin Banks*.

The alphabetical method used here is adapted to the Banthrico field and is a first. It will not apply to cast iron, tin, and ceramic banks. These rating levels reflect the availability of the Banthrico banks in the market place as follows:

A- Very easy to find. One or several can be found at almost any toy or antique show/shop. May even be found at garage and yard sales and general auctions. A generic type bank. Approximately 25 to 50 thousand were made, but many did not survive the ravages of time.

B- Slightly more difficult to find, but still plentiful at toy and antique shows/shops. Available at toy and bank auctions. A generic type bank. Approximately 15 to 25 thousand were made, but many did not survive the ravages of time.

C- Becoming difficult to locate. May have to visit several antique and toy shows, antique and toy dealers/auctions or a private collection before you acquire the bank you are wanting. Usually a specific type bank. Approximately 5 to 15 thousand were made, but some did not survive.

D- Difficult to find. Rarely found at antique and toy shows. Usually found at toy or antique auctions or in a private collection. A specific type bank. Approximately one to five thousand were made and the survival rate is greater because of the low number made.

E- Extremely difficult to locate. Rarely found anywhere except in toy auctions, museums, and in private collections. Issued in a very limited number. A specific type bank. Approximately one thousand were made and most of them have survived due to limited production.

F- Rare. Only a few are known to exist in private collections or museums. A special order or pattern bank or the issue was destroyed except a few pieces. A specific type bank. Approximately 100 were issued or survived.

Generic- Stock banks made by Banthrico and used by many organizations through the years to advertise their company or organization. Examples: Bust, transportation, animal banks.

Specific- Special order banks made by Banthrico for one company or organization. The molds belonged to the company/organization ordering the banks and they sold their banks to the public. Examples: mascots, architectural, trophy banks.

Exceptions- There are exceptions in both generic and specific type banks and the ratings will reflect that in the picture captions.

Numbers

The *numbers* given in each rating are "ball park" figures at best and based on an estimate of the number of different banks made by Banthrico and the banks that were issued from them over a 50+ year period. Many Banthricos did not survive because of their fragility and the rating levels have taken that factor into consideration. Another factor taken into consideration in rating the banks, which is often overlooked, is the number of banks that are "firmly lodged" in private collections around the world outside the bank, building, model car, political and other collecting fields. These two factors reduce the actual numbers available in each level considerably. There are also other factors involved in rating the banks such as age of bank, number issued (records), scarcity (number seen or reported), and type of metal (white metal or lead). There are also extenuating circumstances in rating the banks such as a very limited issue (250 or less), banks were destroyed or melted down, it is a pattern bank (and may not have been issued), or the majority of the banks were shipped to another country.

Trap Types

The rating given on a particular bank also includes the *trap type*. If the same bank is issued several years later with a new trap type, then the rating would probably change. For example, a bank has a T1 trap and is rated a "D" bank and is issued several years later with a T4 trap, then the bank would usually be rated a "C" or "B." If the same bank is then issued again several years later after the second issue and has a T10 trap then the bank would usually be rated a "B" or "A" bank. It would not be fair to place a "D" rating on a bank made in the 1980s while its counterpart made in the 1950s carries a "D" rating. The 1980s bank is much newer and has a new and different trap.

It is obvious that rating the Banthrico banks is not an easy task. The author selected several Banthrico collectors/experts (12) to serve on the "Banthrico Board" and rate the banks. Those ratings with the author's ratings plus input from other sources and collectors were used to arrive at the rating of each bank.

Banthrico made several thousand Lincoln bust banks which would easily qualify it as an "A" bank, and on the other hand, only a few Thomas Dewey bust banks are known to exist which qualifies it as an "F" bank. This illustrates the wide range of the number of banks made and the difficulty of rating those banks in between the two extremes.

Is the approach foolproof? No, but it is a "giant step" in the right direction. As more Banthrico bank information and records become available in the future the rating levels may need correcting and fine-tuning, but for now they seem fairly accurate and objective.

Pricing

The pricing of Banthrico banks is closely related to the rating levels of the banks. An "A" bank will usually bring the lowest price, while an "E" or "F" bank will usually bring the highest price. The price ranges that will be given for the different rating levels will be for banks that are in good (G) to excellent (Ex) condition. A separate section addresses the condition of a bank and gives the different grade levels. A bank condition less than good would not bring the lower part of the price range given. A bank in mint condition or mint in a box (MIB) would command a higher price of 20 to 30% above the range. As the bank ratings move up from A to F, the price ranges become greater because of scarcity.

Because of the different major categories of banks made by Banthrico, the pricing of Banthrico banks will fall into five major areas:

(1) regular white metal, steel/wood banks;

(2) paper/cardboard, plastic, chalk, and vinyl banks;

(3) pocket banks;

(4) model car white metal banks with white and black rubber tires;

(5) architectural white metal/lead/steel/plastic banks.

Because of special collectors outside the bank collecting field, the prices of the model car banks and architectural banks are usually higher than the other three categories. The Souvenir Building Collectors Society (SBCS) is a good example. Its members specialize in collecting all types of buildings — banks, inkwells, doorstops, pencil sharpeners, souvenirs, paperweights, jewelry boxes, etc. Therefore, supply and demand are heavily impacted which pushes the price of building banks upwards. Both the Still Bank Collectors Club of America (SBCCA) and the SBCS members are vying for the building banks.

The price ranges:

(1) **Regular White Metal, Steel, Wood Banks**

		Examples
A	$10 - $25	People, Mascots
B	$25 - $50	Fictional Characters, Animals
C	$50 - $75	Household and Food, Trophy
D	$75 - $125	Calculating, Registering, Misc.
E	$125 - $250	Transportation
F	$250 - $500	

(2) **Paper/Cardboard, Plastic, Chalk, and Vinyl Banks**

A $5 - $10
B $10 - $20
C $20 - $40
D $40 - $60
E $60 - $80
F $80 - $100

(3) **Pocket Banks**

A None Available at This Level
B $85 - $100
C $100 - $125
D $125 - $175
E $175 - $225
F $225 - $500

(4) **Model Car White Metal Banks with White and Black Rubber Tires**

A None Available at This Level
B $100 - $125
C $125 - $150
D $150 - $200
E $200 - $300
F $300 - $500

(5) **Architectural White Metal, Lead, Steel, or Plastic Banks**

		Examples
A	$ 25 - $50	Buildings, Monuments
B	$ 50 - $100	Statues, Towers
C	$100 - $150	Tunnels
D	$150 - $200	
E	$200 - $300	
F	$300 - $500	

The Valley Casting Company, who took over Banthrico production, is still making all types of banks: animals, cars, buildings, mascots, etc. Many of them are reissues of the same banks that Banthrico made. They purchased the bank molds when they bought out Banthrico. They are also making new banks never made by Banthrico. Therefore, their banks are rated lower and priced accordingly because the number that will be made has not been finalized. An organization or company may order several more lots of 500 or 1,000 banks from Valley Casting before they are through. Chapter 14, on Valley Casting presents more about their company and the banks they are marketing.

There are secondary factors that enter into the price of a Banthrico bank besides the primary factors of the rating and the condition of the bank. Some of those secondary factors that impact price are specialty collectors, bank is needed to complete a collection, supply and demand in a particular geographical area, lack of knowledge by the collector regarding the rating and condition of a particular bank, the collector's financial resources, and the dealer or seller's inventory.

The pricing of the Banthrico banks is based on the experience of several advanced Banthrico collectors, including the author. These collectors are mainly located in the Midwest, but have purchased, sold, and traded banks with other collectors in all geographical areas of these United States, so the price ranges should be fairly accurate. In the final analysis, the price of a given bank is what the collector is willing to pay for the bank and what the owner is willing to charge for the bank. If a collector wants a particular bank and the price is too high and the seller will not budge on his price then one of two things will happen. The collector walks away from the bank for sale or he buys the bank. If he buys the bank, that is "the price" of that particular bank at that particular moment in time. A willing buyer and a willing seller have just "set the price" in the market place. Most bank collectors have paid too much for some banks in their collection, but they have also found some bargains for their collection. Some banks purchased were above the price range and some banks purchased were below the price range. That all seems to be part of the fun, challenge and adventure "in the hunt for those wonderful banks."

The price ranges given for the different rating levels is purely a guide to help the collector and dealer know what is the usual charge for a given bank. The price ranges are not engraved in stone and are intended to be flexible. The price the dealer had to pay for the bank will impact his asking price and the knowledge and wallet of the collector will govern his decision to buy or to pass. I have no doubt that as we know more about the numbers of different Banthrico banks that have survived and are available that the prices will need adjusting. It may be up and it may be down. My educated guess is up.

Condition and Grading

The condition of a Banthrico bank is very important to a collector. Most collectors want to have the best condition they can afford and is constantly upgrading the collection from fair or good to excellent or mint. As the collector climbs the "condition ladder," he acquires trading material for other banks and the upgrading process is repeated with other collectors. Some collectors are satisfied with a bank in good condition, others want nothing less than fine condition and some will settle for mint only and even mint in the box (MIB).

If a Banthrico is difficult to find (C, D, or E), then the collector is willing to move down the condition scale to acquire a given bank. On the other hand, if a bank is readily available and an A or B bank, then the collector becomes more discriminating and wants a mint bank or MIB bank. He will continually upgrade his collection of A and B banks until he achieves that goal. Even though the collector would like to have a C, D, or E bank in excellent or mint condition, the price might prohibit its acquisition.

The condition of a bank has a direct relationship with the price. The better the condition of the bank the higher the price. You will note that in the pricing section, a price range is given for the different ratings of the banks. Those in the lower price range are for banks in good condition while those in the higher price range are for excellent specimens. Those that are mint and MIB actually move out of the price range and command a premium. Those banks in fair condition will not usually bring the price given in the various price ranges by ratings because a fair rating is below a good rating.

The bank's condition impacts the collector on what he can afford. If an A bank costs $30 in excellent condition and a D bank costs $125 in excellent condition, the beginning collector may choose the A bank because of the price. If a seasoned collector needs the D bank, he will probably purchase it. If an A bank in excellent condition is $40 and a C bank is in good condition for $50, the collector has to weigh condition, price and rating. Often the collector will choose to purchase the C bank because of its rarity and pass on the better condition A bank. Other collectors would choose the A bank because of the excellent condition and it fits the wallet better for that day. That's what makes bank collecting so much fun and an adventure — first the hunt, and then the find, and last, but not least, the decision — what to purchase or when to pass.

So **the ratings, the price, and the condition are all intertwined**. The rarity (rating) of the bank regulates the price ranges and the condition of the bank moves the prices up and down the ranges within each rating level. The rating and pricing sections need to be reviewed together with this section to help understand their relationships.

Some collectors will not purchase a Banthrico in fair condition no matter how the bank is rated. Other collectors will pick up the bank as a filler in their collection until they are able to upgrade the piece. There is a direct ratio in keeping a bank in fair condition to its rating level. If it's an A or B bank, most collectors will pass on the bank knowing that they will soon spot another one in better condition. If it's a C, D, or E bank, there is a good chance that the collector will buy the bank if it's not in his collection. Collectors will not usually buy fair banks for trading material because they are not that desirable and difficult to move. An investment of $20 to $60 in a fair condition bank for trading is not usually a wise investment (unless you know a collector who needs the bank and will pay you for it).

The condition of Banthrico banks need to be divided into two major areas. The first major area is the white metal, lead and steel banks. The white metal banks were the major product line and the greatest contribution of the Banthrico Company to collecting and that needs to be the main focus in rating, pricing, and grading a bank. Lead and steel banks were produced early in their history, but not in the quantity and the different banks that were made in white metal. The second major area is the plastic, chalk, paper/cardboard and vinyl banks. These group of banks show different signs of conditioning and wear and will be treated separately in the book.

Not many of the white metal Banthricos were painted, but that factor will be included in the grading/condition of the bank. The Sleeping Santas, the Seated Indian Boy, the Boulevard Bank building and the autocars are fine examples of painted white metal Banthricos.

The majority of the white metal banks were lacquered after they were made and this process enhanced their life span. Zinc and aluminum, the major alloys in the white metal banks, are highly susceptible to oxidation, corrosion, and discoloration. They are very vulnerable to wet and damp areas. Just leave a white metal Banthrico in a damp basement for a year and you will see what I mean. They need to be kept in a warm and dry atmosphere. The lacquer coating protects them from the elements (salt, acid, etc.), but the elements eventually wear through in areas with a thin lacquer coating. When you see dark tarnished spots on the Banthrico white metal banks, that is where the lacquer has failed and minor oxidation has taken place. If the banks are kept clean and dry, the oxidation process is slowed down considerably. When you see a Banthrico with a heavy white discoloration and sometimes holes eaten into the metal, oxidation is in "full bloom." The following section on cleaning and caring addresses what can be done to protect your collection. The amount of dark tarnish spots and degree of oxidation on a bank impacts its condition and price.

The fragility of white metal banks is another factor that impacts the condition of the Banthrico. The white metal banks have very thin molds and walls and are easily dented, bent, dinged, and scratched. Banthricos with no damaged areas increase their grade level considerably and bring higher prices than those that are damaged. A Banthrico in fine condition will not have any dented areas, yet one in good condition might have some minor dings, and a bank in fair condition might have some major dents and bent areas. The white metal banks need to be handled carefully and not bumped against each other. Most collectors are disturbed when someone is viewing their collection and the person constantly strikes the banks against other banks. When shipping white metal banks, they need several layers of bubble wrap to protect them.

Missing parts on a white metal bank will also impact the condition. When the trap is missing or broken, this has a significant impact on the rating. A bank in excellent condition without a trap will pull that grade down to fair. If the trap is bent and broken but still intact on an excellent bank, it will be graded in good condition. Missing keys on key lock trap banks do not impact their condition of a bank, but a key with a bank does enhance the price by a few dollars ($2 - $10). The reason I make that statement is because the different keys are easily obtainable and interchangeable with many traps. Keys will be discussed in the chapter on traps. As we all know, keys are easily misplaced and lost. Banthrico did not make one special key for one special bank — they are generic. Even specific banks had generic keys. Therefore, they will not be included in the grading levels. There may be other missing parts on the Banthrico bank and depending upon it being major or minor, will impact the condition in varying degrees. The flag missing on the Astronaut bank would certainly be minor and have little impact on condition, but the pocketwatch missing on the clocktower bank would be major and bring an excellent condition bank down to good or very good condition. The same would be true of the black rubber tire autocars. If the wheel(s) are missing or replaced with other wheel(s), then the condition has a major impact and would be reduced accordingly.

The condition of a bank will *not* affect the rating of a bank. A bank is an A-B-C-D-E-F bank based on its *rarity* and not on its *condition*. It may be an E bank in fair condition, but the condition does not change its rarity. On the other hand, the condition of a bank *does* directly impact the *price* of a bank, as stated earlier.

Repainting Banthrico white metal banks will reduce the grade or condition of the bank. Many of the autocars with black rubber tires have been repainted and done in a very professional manner. Many of the autocars have little paint remaining and that encourages the collector to repaint them. Several of them are included in this book to show the different colors. But the repainted bank has to be moved down a level or two in grade depending on other conditional factors of the bank. A repainted bank cannot be graded the same as the bank in its original color. Sometimes you will find a painted lacquered bank that was not originally painted by Banthrico, but has been painted by an enterprising bank collector. You will see one in the antique auto section of the book just to show the variations that exist. Painting an unpainted Banthrico bank will reduce the grade and price considerably.

The condition and grade of Banthrico banks:

White Metal, Lead and Steel Banks

Fair - Banks in this condition may be bent, have pieces missing, sometimes have holes punched in them, the trap is missing or broken/bent, it usually has heavy oxidation/corrosion and heavily tarnished. The bank is identifiable and some of the writing can be read. There is less than 50% of the original paint. Any single or combination of the above characteristics would grade the bank as fair. The collector will purchase this bank only as a filler until a better one shows up. Occasionally, a D or E bank is found in this condition and the collector may clean it up and put it on his shelf and be satisfied with it, but that is not usually the case. Probably about 10% to 15% of the banks available fall into this category.

Good - Banks in this condition have all their parts, have no bends or punctures. They might have a small ding. All the trap is in place, but might have a bend on it and may not work properly. There may be slight rust on the trap, but not heavy rust. There may be partial or total dark discoloration and tarnish of the bank. There may be very minor surface oxidation in small areas, but no corrosion. Overall, the bank looks good and is collectable. Most (60% to 70%) of the advertisement can be read on the bank (if there was any) and all embossed/raised/recessed words are easily read. There is at least 50% to 75% of the paint. Probably about 40% to 50% of the banks available are in good condition.

Fine - Banks in this condition are bright, shiny and well lacquered or have 75% to 85% of their original paint. There may be an occasional small dark spot on the bank. There is no oxidation or corrosion present on the bank. There is nothing missing and the trap is not broken or bent and is in proper working condition with no rust on it. The bank may show only slight wear in a few high spots. All writing and embossed wording/symbols are easily read. The bank is highly desirable and looks great in any collection. This is the condition level that Banthrico collectors strive to achieve. They may get a better conditioned bank, but they will usually be satisfied with a bank in fine condition, especially if it is a C, D, or E bank. Probably about 20% to 25% banks available are in fine condition.

Excellent - Banks in this condition are also bright, shiny and well lacquered or have 85% to 99% of their original paint. You can tell they have been handled and circulated, but cared for properly. There are no dark spots on the lacquered bank and there is no wear on the high spots. There will be no oxidation or corrosion on the bank. The bank will have no missing parts and the trap and its mechanism are fully functional and just as excellent as the rest of the bank. All writing and embossing are in top notch condition. The bank may be in a box and have the key, but that does not necessarily make it mint. It may be in excellent condition. The mint grading criteria will determine if it achieves that level. Probably about 5% to 10% of the banks available are in this condition. It is difficult to find a bank in this condition, but when found the collector will gladly place this gem on the shelf with his other banks.

Mint - Banks in this condition have no indication of handling or having been in circulation. They have 100% original paint, they are often found in the box with the key (Mint in the Box - MIB), although the box and key are not required for a mint grading. They have that "new luster" finish about them that is a step above excellent. The bank is as near perfect as a bank can be. There can be no imperfections of any kind at this level of grading, including a paint chip. Many collectors try to put excellent banks in this grade level, but it is in a class all by itself. There are not many banks that qualify for this pristine grading. Probably around 1% to 2%.

Two Extra Levels

To help people in grading banks and arriving at an agreeable price, two additional levels can be added to this grading. The *first level being very good*, which would be halfway between good and fine, and the *second level being very fine*, which would be halfway between fine and excellent. By having these two extra levels, it gives the seller a chance to expand the grading if he feels his bank is slightly better than good, but not fine. Therefore, it is very good. The same grading approach being true for very fine. Grading at these two mid-points of very good and very fine need to have characteristics that are distinguishable and discussible between buyer and seller that clearly shows the bank is in slightly better condition than the lower grade in question. This is where the "big challenge" comes in agreeing on the grade and the rating and arriving at the price for the bank.

Please note: *on grading the lead banks*, they are darker by nature and may not have the bright and shiny appearance that white metal banks have. That factor needs to be considered in grading the bank, but all other factors listed on grading hold true.

Plastic, Chalk, Paper/Cardboard and Vinyl Banks

Fair - Banks in this condition are heavily worn, frayed, "dog-eared," and chipped. Have less than 50% of their paint and may be dented or punctured. Usually just a "filler" for a collector. May have some parts missing.

Good - Banks in this condition have no dents, punctures, or missing parts and have 50% to 75% original paint. May be a few dings and paint chips and the majority of the advertisements on the plastic and vinyl are readable. There are no major tears on the paper/cardboard and the advertisement is fully readable, but will be worn. Overall, the bank looks good and is collectable.

Fine - Banks in this condition are colorful and shiny. A few minor scratch marks may appear on the plastic, chalk, or vinyl and a wrinkle or crease on the paper/cardboard. They have 75% to 85% of their original paint and only slight wear on the high places. The collector is pleased with the overall condition of the bank and will gladly add it to his collection.

Excellent - Banks in this condition are showcase pieces with 85% to 99% original paint. The plastic and vinyl are "like new" and there is no wear on these banks. The paper/cardboard is bright, well printed and has been handled very carefully and very little.

Mint - Banks in this condition have no indication of handling or having been in circulation. They have 100% of their original paint. They may be found with the original box and key, but they are not required for the bank to be mint. The bank is clearly better than excellent and is as perfect as a bank can be. There are no imperfections of any kind at this level of grading. If there is an imperfection, the bank should be graded lower — probably excellent and maybe fine. The bank is unquestionably pristine in condition.

Care and Cleaning

If you properly care for white metal banks, you will go a long way in preventing "the cleaning process." Banks that have not been properly cared for should be cleaned to enhance the bank. Some simple guidelines are offered.

- Dust banks every 3 to 6 months with a soft, warm, damp, cotton cloth and let them air dry. The damp cloth will also reduce the dust level.
- Eliminate excessive handling of the banks, especially with "sticky and dirty hands" (like children have). My banks are mainly in open shelves. I like to handle the banks and I like collectors to be able to handle the banks. But after a bank meeting, when the banks have been handled a lot, I will wipe some of them with a soft, warm, damp, cotton cloth and let them air dry. Some collectors I know have most of their collection in glass display cases and this eliminates handling the banks. If that is the situation, it is good "bank etiquette" not to ask to handle the banks unless the owner makes the offer. The acid, salt, sugar, and other chemicals on hands will be transferred to the white metal banks and, over a period of time, unchecked, will cause oxidation. Oxidation is the biggest enemy of white metal banks; they appear as green, gray and white spots. When chemical substances get on the bank and are combined with oxygen, they act as oxidizing agents and bring about oxidation on the bank over a period of time.
- Don't leave your banks in a greasy or oily area (like the kitchen) because these substances will form on the banks and eventually discolor them and cause oxidation.
- Prevent rough handling of the banks such as banging and striking them against each other. The walls of the banks are thin metal and they will bend and puncture easily. White metal banks should be handled carefully to insure longevity.
- Place your banks in indirect sunlight or lighting. If left in a very sunny location over a period of time, there will be some fading and discoloration depending upon the depth of the metal and lacquer coating. Even bright electric lights shining directly on banks over a long period of time will cause some fading.
- Keep your banks in a warm dry area. If they are stored, it is better to place them in the heated section of the house rather than in the basement or garage, which is often unheated or damp. The attic is a better storage place for white metal banks than the basement or garage because it is usually warmer and drier. If your white metal banks are kept in the finished basement of your home, a dehumidifier is recommended.
- When shipping your banks, be sure to pack them properly. They should be well wrapped with heavy bubble wrap and then fully lined with Styrofoam popcorn or crumpled newspapers to prevent the banks from moving around in the container. A sturdy cardboard box should be selected for the banks journey (not a thin-walled cardboard box). Be sure to insure the banks for their proper value. I know several collectors that have failed to insure their banks and they were lost in transit. We all know that packages are handled roughly by someone

along the line, so pack accordingly. Don't let someone else pack your banks, do it yourself or supervise the packing.

When it comes to cleaning white metal banks, there are many different opinions. There is almost a different opinion for every collector you talk with about cleaning banks. There are some purist collectors that totally refuse to clean or dust a bank in any manner. How they found the bank is the way it will stay. There are other collectors that will dust the banks and that is about the extent of their cleaning. Some collectors tackle minor cleaning of a bank and then there are those who attack a bank "big time" if it needs cleaning. Somewhere in one of these groups you will find agreement or maybe in-between one of these groups. I am probably close to the last group. I like clean banks and want them to look their best. Yes, I have made some mistakes in cleaning banks, but the more you experiment and the greater the experience, the better the success rate.

If you decide to clean your banks, it is best to start with one that is inexpensive and maybe even in fair condition. It might even be a wise move to purchase an inexpensive, fair condition white metal bank to use for practicing your cleaning methods. The different cleaning methods can usually be tried out on the same bank to see the results. If you mess up, the loss is minimal. Those who have never cleaned a bank and start on an expensive bank is a bad combination and usually a regretful scene.

I have had my best results in cleaning white metal banks with WD-40. It can be purchased at any hardware or auto store. It is an oily cleaner that dissolves dirt, rust, grime and goop. It does not bother the metal, lacquer, metal coating or paint. I spray it on in a large metal pan and let it soak for 2-5 days, depending upon how dirty and oxidized it is. I spray it each day to assure it has plenty of WD-40 on it. If there are real bad spots or dirt in the details of the bank, I brush them briskly with an old toothbrush before spraying it. This process is repeated several times daily until you are satisfied with the results. The bank is then wiped dry with a soft, dry cotton cloth to remove all the WD-40. It is best to set the bank on a fresh piece of dry cotton cloth for a few days to get any residual WD-40 that may drain off the bank. I think you will be pleased with a much better looking bank in your collection.

Cotton buds (Q-tips) and round toothpicks are great for cleaning out the corners, crevices and details of the banks. I have had good luck with Soft Scrub cleanser for those heavily oxidized spots. It is tough on greasy dirt, mildew stains, oxidation/corrosion, and scum buildup. It is rubbed gently with a damp cloth or sponge on the areas needing attention. It is also good for the steel and aluminum traps on the white metal banks and can be used a little more aggressively since the surface is harder than the white metal. Wipe dry after each application and repeat the process until the area is acceptable. If the bank has some very rough oxidation areas on it or heavily rusted on the trap, 4-0 (0000) steel wool (very fine) or Brillo pads (steel wool with cleanser) can be used, but be careful not to scratch the surface. Sometimes it is difficult not to scratch the bank and trap so you will have to decide which looks best — the oxidation/corrosion or a few scratches. I prefer the few scratch marks rather than the oxidation and rust. You also have stopped, or slowed down, the oxidation and rusting processes by cleaning the bank. It is up to each collector to decide the extent they want to clean their banks. This process usually improves the looks of the bank and has helped me in the value of my banks for selling or trading. The more expensive the bank, the less likely the more vigorous cleaning processes should be used. Anytime you clean a bank there is an element of surprise and the unknown, so please keep that in mind. The more experienced you become in cleaning banks, the better the results will be.

When you need to remove old or new price sticker tape and other sticky substances from your white metal banks or steel/aluminum/tin/plastic traps, vegetable or corn oil from the kitchen will do the trick. You may have to soak it for a day or two to loosen up the gum and goop, but keep scratching it with your fingernails and it eventually will all come loose. The oil engulfs the sticky gum and it has no adhesive power.

Some of the traps on the Banthrico white metal banks can be removed fairly easy with a screwdriver. Some of the traps are soldered down, some braided on and some have screws that are machine installed and cannot be removed without destroying them. If the traps can be removed with a screwdriver, it is wise to clean both sides of the trap if the outside needs cleaning. Use the same cleaning procedure on the inside of the trap as the outside.

If the inside of the bank is oxidized and/or dirty, the WD-40 can be sprayed on the inside and follow the same cleaning procedure as the outside. It is much more difficult to dry the inside of the bank with a soft dry cloth and more draining time will be needed for the inside of the bank. The WD-40 spray can comes with a 5"-6" plastic attachment that is excellent for spraying the inside of the white metal banks. The inside of the bank does not usually get the wear and tear and dirt that the outside of the bank receives, so unless it needs special attention, I do not touch the inside.

It is best to under clean a bank section than to over clean it because you do not want the cleaned bank section to stand out, but to blend with the rest of the bank. If you do too much cleaning in one area, you may wear off part of the lacquer and metal coating and the bank will look worse than before you started. If you have a bank with a worn or bare spot, you might try Rub-n-Buff metallic matching paint to cover the area. It can be purchased at your local hardware store. It is easy to use and remove. Just follow the instructions that come with the paint.

If I secure a white metal bank for my collection that just needs a good bath, I use Liquid Dove or Ivory Flakes in lukewarm water. After the thorough washing I use a hair dryer to speed the drying process and insure the inside is also dry if it got wet. Some collectors do immerse their banks in water, but I do not as a general rule. If there is still some stubborn dirt on the bank, a little Fantastic rubbed with a medium bristle brush is another cleaning tool than may be used. Wipe with a clean damp cloth immediately and use hair dryer.

Often, white metal banks end up with paint on them for one reason or another which, in most cases, is carelessness. These paint spots can be removed with a Q-tip and some fingernail polish remover (acetone). Put a small drop of remover on a Q-tip and rub it on the paint spot. It should come loose rather quickly. Wipe the area immediately with a dry cloth to prevent invasion into the lacquer and metal finish. If the paint will not budge, try removing it with your fingernail or a wooden popsicle stick. Do not use a knife blade because it will usually scratch the finish.

Other collectors have used lemon juice, vinegar, ammonia, naval jelly for rust (traps), white metal polishes and refinishers (found in hobby shops), polishing cloths, Nevr-Dull (the magic wadding polish), and preservative oil CRC5-56 to clean their white metal banks. I have not tried these methods, but other collectors swear by them as I do by my cleaning methods.

After the bank is cleaned to the degree that you like it, apply a thin coat of Johnson's colorless paste wax (or any brand that you favor, and it can be paste or liquid wax) with a soft, clean brush or cloth. Let the wax set for 30 minutes and wipe off excess wax with a dry, soft cloth, clean out the crevices with a Q-tip, buff the bank with a clean shoeshine brush or lamb's wool shoeshine cloth. You will be delighted with the results! Who said you should not clean banks? Proper cleaning will enhance the bank's appearance and increase its value.

Some closing thoughts on the care and cleaning of white metal banks:

Don't hurry the process. Take time to clean the bank properly. Be patient.
Don't use hot water.
Don't use strong chemicals and abrasives.
Don't clean a bank if it doesn't need it or if you are satisfied with it. After all, it is your bank to enjoy.
Don't start your cleaning experience with an expensive bank.
Don't expect the bank to be in excellent or mint condition when you finish cleaning it. Expect it to be improved.
Don't leave your banks in a damp, moist area.
Don't ship your banks until they are properly packed.
Don't mishandle your banks.
Don't over clean your banks.
Don't forget to seal your bank after the cleaning process with a thin coat of clear wax to protect it.
Learn from the experience and do an even better job next time.

Most white metal banks that we desire to clean are in the fair to good condition levels. When a bank reaches the fine condition level, I do not recommend cleaning the bank, only dusting it.

Manufacturing

The author met with Peter Eisendrath, the last Banthrico president, to discuss how Banthrico made their white metal banks. This was the "heart and soul" of the Banthrico Company's work. Questions had been prepared and the answers he gave, plus additional comments and research by the author, tell this part of the story.

This information is presented in simple language and does not go into great detail. Bank making is a highly complex art with many steps and would require extensive development to cover the subject fully. Many highly skilled people are involved in the process. The management planner, sculptor, die (mold) maker, engraver, caster, chemist, metallurgist, industrial engineer, quality control inspector, packer, shipper and many others are involved in the bank making process.

What was the metal composition of the white metal banks?

They were mainly zinc (95-97%) with 3-5% aluminum, 1% or less copper and some other trace metals from the ore. The manufacturer puts copper in the metal because it helps in the pouring process. The various zinc and aluminum percentages may vary slightly by 1% or 2% because of the different alloying from the different smelters. (A smelter is a company that digs the raw ore out of the earth, heats it, melts it, and separates the various metals into ingots (bars) of metals. Smelting is a refining, reducing, separating process of the metal ore. An ingot is "a mass of metal cast into a convenient shape for storage and transportation to be later processed" according to Webster's New Collegiate Dictionary. Zinc is a bluish white crystalline metallic element that is abundant among the earth's minerals. It is easily heated, combines well with other metals, has a low to intermediate hardness and is not costly compared to other metals which makes it ideal for bank making. When the zinc, aluminum, copper and other trace metals are mixed they become an alloy.)

Where does the metal come from and how was it transported to Banthrico?

The zinc ingots came from several states all over the United States. They were usually brought in by trucking companies contracted by the smelter. Ten thousand pounds of ingots were delivered at a time. Banthrico might use most of the 10,000 lb. of ingots in one week if we were exceptionally busy. The ingots weighed 50-100 lb. each but could be broken off easily at the various break points (narrow sections) to make handling easier.

What temperature is the white metal heated and in what type of container?

The white metal ingots were placed in large, round cast iron kettles that are some three feet in diameter and 12 to 15 inches deep. The kettles had gas burners under them for the heat with a hood over the top to take out the fumes. There were thermocouples (electric measuring instruments) in the kettles to regulate the metal temperature. The temperature of the liquid white metal was 625 to 675 degrees Fahrenheit (note: the boiling point of water is 212 degrees and we usually bake potatoes and chicken in the oven at 350 degrees to give us an idea of just how hot the white metal is when it is liquid). The kettles were on stands with metal bases around them. The floor was wood beyond the metal base covers so fire safety was of utmost importance.

How are the bank molds made?

First, we start with a photograph, blueprint or an artist's rendition. The sculptor made a plaster model of the actual size of the proposed bank. Sometimes they used wax, but usually plaster. We had one artist (Alfred Brunettin) that worked for Banthrico as a sculptor from the late 1940s through the 1970s that did the largest majority of the banks — almost 30 years (Brunettin's signature appears on some of the John F. Kennedy busts - left side). Lois Mihok did some of our banks in the 1970s and 1980s. The plaster model was sent to the customer for their approval. If any changes were needed, they were made by the sculptor. Sometimes some of the details of the bank had to be eliminated to make room for the bank name, coin slot, etc. Then a plaster shell was made from the plaster model and sent to the foundry to make the metal mold. Some models were wood, such as the eagle, ram, pelican on pier posts, and cougar.

The foundry would make a brass mold (because they last longer and fit together better) from the plaster shell. The brass mold had some three to six pieces on the average. Some more elaborate molds might have seven or eight pieces. The mold pieces were fitted together at the bottom with individual clamps and could be separated individually when the bank was removed from the mold. For example, if a building bank was being made, it would probably have four sides and six (counting top and bottom and four sides) pieces to the mold. The bank was filled with molten white metal with the top of the building on the bottom of the mold. The gate is the piece of metal at the top of the mold (like a collar) with an opening where the molten metal enters.

The separation of the mold pieces becomes very important in the bank making/molding process. The mold seams have to be tight so they test the mold with lead. If the lead seeps through the mold, the mold pieces are ground down and refitted. If the lead does not seep out, the mold is ready for use.

Banthrico had two tool and die makers that worked with the foundry and kept the molds working properly after Banthrico received them from the foundry. The bank mold was placed in a moveable cradle with the top of the bank on the bottom and the bottom of the bank on top. The moveable cradle allowed the bank caster to pour the excess metal out of the mold. The molds were the property of Banthrico, but the company that had a specific mold/bank made had exclusive rights to the mold. They could not be used for other companies. When Banthrico sold their company to Toystalgia, the molds went to Toystalgia. Pete remembers his dad, Joseph, selling for scrap metal old bank molds because of a space problem (not realizing their future collecting value).

What is the engraving procedure?

The engraving is done in reverse or backwards so that when the mold is filled with liquid metal, the spelling comes out correctly on the bank. When letters are engraved into the metal mold they come out as raised letters when the mold is cast (filled full of molten metal) because the molten metal fills up the recessed letters. If there is a mistake in the engraving or the customer changes their mind on any of the wording, a "pocket" is made in the metal. The metal is ground out and filled with metal and the engraving redone. The Wrigley Building was a good example of a bank that was redone due to the Wrigley management wanting the wording changed. The increased cost was paid by the customer requesting any changes from the original contracted bank. Engraving is an expensive art.

Author's observation: This is the reason why many of the building banks did not have the city and state on them which is very frustrating to bank and building collectors. It is extremely difficult to identify the building banks and other banks when all the information is not on them. The customer was only interested in his "own little world" and the cost and had no idea that forty of fifty years later collectors would be wanting this information on the bank. Who would have ever dreamed that a white metal bank made in the 1940s through the 1960s for $1.60 would ever be worth $160.00 or more today. That is 100 times the original cost! What an increase in cost and value! If we had only known!

How did the caster (those who cast the bank into the mold using molten white metal) get their education and training and how did they dress?

They learned by experience, trial and error, and on-the-job training. If we had an employee who showed an interest in this area and met the intellectual and physical requirements, we trained them. Casting is a very demanding job that requires alertness, insight, analytical ability, physical stamina, and strength. We had one man who was a caster for forty years. He was like a machine. He rarely stopped casting banks to eat, drink, or go to the bathroom. The casters did "piece work" and were on commissions so the more banks they made, the more money they made. We had four or five casters working all the time and had little turnover. Each caster would be working on a separate bank. One caster might be casting a building, another a bust, one an animal, and another a mascot or car. The caster wore gloves, a heavy cloth apron, heavy leather shoes with steel toes and goggles to protect his eyes from the fumes and hot liquid metal. The gloves, aprons and shoes of the casters had small burn marks where the hot metal droplets landed while they were working.

What is the amount of white metal per bank?

The caster left 14 to 16 ounces (average) of molten white metal in the mold. The bank usually weighed close to one pound. If the bank weighed more than it was supposed to, it was rejected (more than two ounces). An inspector checked (weighed) the banks and if rejected, they were placed back into the kettle of hot molten white metal and melted down. The caster only got credit for banks that were accepted.

Joseph Eisendrath (your dad) said that "hand casting was an exciting art and we had high standards." He also stated that the "casting business is the core of our operations." Will you walk us through the hand casting process of the caster?

The casters worked between the kettles of molten white metal and the bank molds sitting in a moveable cradle-like holder. They have a ladle (dipper) that is 6 to 10 inches wide and will hold several pounds of molten white metal. He dips the ladle into the kettle and slowly turns and pours the 650 degree metal into the mold. He rocks the molten metal back and forth for five to ten seconds in the mold (on cradle) and then tips and pours the excess molten metal in the center of the mold back into the ladle. The ladle is dumped in the kettle and placed by the kettle. White metal is left adhering to the sides of the mold. He immediately takes a rawhide mallet and knocks the mold pieces from the metal cast bank. Each mold piece comes off individually. The bank is still very hot metal. It is placed on a wooden tray right side up. The whole process of pouring, rocking, pouring out excess metal, returning it to the kettle and removing the mold pieces takes about one minute. As one can easily see, the casters cannot be distracted from their work. They must stay focused. The process is called "slush molding." If you left the metal in the mold for an extra minute or two, it would become a solid metal mass. If that happened (and we know it probably did occasionally), the metal would be re-melted in the nearby kettle. If the metal is left in the mold beyond the five to ten seconds, the walls get heavier and stands a good chance of being rejected. Four casters can usually cast approximately 100-200 banks (4 or 5 piece molds) per day (the more complex banks about 75-150). The average bank weighed about a pound, so that is a lot of weight and casting. The caster would clamp the pieces of the mold to the bottom to put the mold back together. Each mold piece has a clamp on the bottom to hold it in place. Once the pieces were put back together and the gate (collar) placed at the top, the caster was ready to cast the next bank. The whole casting operation is a smooth one taking about one minute and another minute or two to reassemble the mold for the next casting. This is what is meant when one speaks of "hand casting" — done by hand, done by man and not by machine. If casting is done by machine, that is called die casting. Banthrico used a zinc metal material called, ZAMAC, in die casting for some of their traps and keys. The metal bank at this stage has a white zinc appearance with no finish on it. There are a couple of pictures in the trap and variation chapter that illustrate what the bank looked like after casting and before the color is added.

Did any of your casters or other personnel transfer to Toystalgia?

No, not one person to my knowledge.

When are the mold marks removed and the color added to the banks?

After the banks are cast and have completely cooled, they are checked for any sharp edges (mold marks). We placed our main emphasis upon quality casting and so had little buffing to do on the banks. We had a belt sander and a buffing wheel to deburr and buff these out where needed. The color of the banks was added by the electroplating process. The banks are placed in racks in a special solution. In the solution were wire baskets of brass or copper balls ("anodes"). When the electric current was turned on and passed through the solution, it would deposit the metal particles on the zinc/aluminum casting through positive and negative attraction. We used three finishes on our banks for years. Satin statuary bronze (red copper), old English brass and chrome or nickel silver. When the banks came out of the solution, they were cleaned and rinsed. They were then placed in a black purple solution (oxidizer) and when removed, the black solution adhered to the banks. The banks were buffed and polished and the black was left on the bank in the crevices to highlight the different parts of the bank, e.g., the names, windows, and doors. This highlighting improved the appearance of the bank. The banks were then placed in a "sawdust" material to dry the castings and to soak off any oils and liquids. The banks were filled with the sawdust and rolled around in it. It was reused several times and then discarded. (If you have purchased many Banthricos, you will note that sometimes all the sawdust had not been emptied out of the bank and it falls through the slot or trap.) All plated banks were sprayed with a clear top coating. We spray painted some banks such as the seated Indian, the Wrigley Building, and the Santa Claus banks. We did do some banks in goldtone and pewter in the late 1970s and early 1980s. There was no plating on the bank. It was a raw zinc casting that was polished, dipped in the black solution, wiped and coated with clear alkyd urea. This lowered the cost of making the bank.

Did you use lead in making your early banks?

I am not sure, but, not to my knowledge. Lead was a health hazard and we did not use it in the manufacture of the white metal banks in the 1960s - 1980s for certain. There were also tight restrictions on the use of lead in manufacturing (Pete joined Banthrico when he graduated from college in 1961). The heavy casting could be some early banks where we used a different weight standard for the banks and the banks had thicker walls (longer time frame in the molds). The author thinks Banthrico did a few early buildings in lead due to weight and fine detail casting. They are specified in the architectural chapter.

When is the clear lacquer coating added?

We sprayed the banks after they had been thoroughly dried in the sawdust and wiped clean. We used clear alkyd urea which is a synthetic resin and provides a high protective gloss on the bank.

When are the traps installed and who made the traps?

We made the traps at Banthrico. We had four stamping machines and made all the steel traps. We also made the steel keys with the stamping machine. The die cast locks and keys are made of ZAMAC (zinc) and we made those by machine. They were made in sets together (key and lock). We installed the trap screws by hand in the 1930s - 1940s, but starting in the 1950s, the majority of the trap screws were machine installed with an automatic screwdriver. The banks were laid in a box holder upside down. The worker drilled the screw holes and then installed the trap plate with the automatic screwdriver. The screws are "one way screws" as most Banthrico collectors know. They go in one way, but very difficult to remove the other way. (As a collector, I wish that Banthrico had stayed with the traditional screw so the trap could be removed, cleaned, changed, etc.)

Explain the silk screen method and any other printing that you used on the banks.

The silk screen method used a piece of treated synthetic fabric with the copy and graphic in the fabric to be used on the bank. The untreated letter areas picked up the ink when exposed, while the treated areas did not accept the ink. We also used rubber stamped printing with ink and hot stamped with gold foil (a heat process) on the plastic banks. The majority of advertising generic banks were silk screened. Banthrico had to constantly evaluate the cost of their bank making process and economize wherever they could.

Did the banks vary in size from one batch to another?

No, they did not. The same mold was used over and over again so there should be no variance in size. What if a piece of the mold was broken and was repaired or replaced? That should have no effect on the size because the mold piece had to fit with the rest of the mold pieces. If a mold was totally destroyed then there could be a variance. The author believes there is a slight variance of 1/16 to 1/32 in some of the banks, especially the autobanks made by National Products and continued by Banthrico as well as some buildings. There would be a slight variance of an ounce or two in the weight based on heat of metal and time in mold.

How were the banks prepared for shipment?

We packed forty banks to a case. Our price was based on 160 generic banks so that was four cases. We had to use boxes that were free of sulfur (acidic content) because sulfur would attack the plating. The Railway Express Agency (REA) trucks handled a lot of our shipments in the 1950s - 1970s. (Be careful that your Banthrico banks are stored in paper and boxes free of sulfur). Later plastic bags replaced tissues to protect the finish.

Did Banthrico make any plastic, chalk, composition, cardboard, wood or vinyl banks?

Not in-house. We contracted out all of these banks. The plastic American Home Bank was done by companies in Aurora and Chicago, Illinois. Riverdale Plastics did our vinyl banks. Japan and Taiwan did our paper mache (composition) banks. Wisconsin Can did our tin trap plugs. We did not do the tin traps on our stamping machines. The "Bozo the Clown" bank was a limited edition plastic bank we contracted out for the National Tea Company. When Toystalgia bought out Banthrico they did not want the cardboard coin folders which we contracted out. We made millions of them through the years. Pete kept the coin folders in his present sign business, Signs Now, and still sells them. The wood barrel banks were first made by the Bankers Thrift Company and Banthrico continued the line for a few years after we bought them out in 1931.

Following are additional comments about the Banthrico Company made by Pete Eisendrath in general discussions during the April, 1999 interview:

All the Banthrico records on the different banks were taken by Toystalgia (Valley Casting) when they purchased Banthrico.

Banthrico made a Canadian classic car set of high quality white metal with an ID serial number plate on the bottom. It was a limited edition of 550 sets, but was not a bank. It was done in the mid-1980s to broaden our line of products.

We made a Henry J car bank and a 1953 Corvette model car (not a bank — one thousand issue).

The 1902 Pillbox was the name we used on that particular antique auto we issued.

Tom Aronson, the Vice President of Sales for Banthrico when we closed, is now a business broker in Chicago, living in Deerfield, Illinois.

The autobanks (black rubber tires) that we made were based on actual car blueprints from Detroit. We also used the same car colors they did for the autobanks.

Banthrico had a National Ronzite Division that made plaques, honor rolls name plates, and all types of brass plates after WWII.

Custom (specific) banks had to be an order of at least one thousand for the first order and the second and following orders could be for five hundred. Seventy-five percent of the custom orders were for one thousand only, but there were others (25%) that placed multiple orders after the initial order.

We had a Banthrico sales force of about 15. They were assigned a "territory" and usually lived in the area. Some territories were half a state like Illinois. Missouri was a territory. Texas, Oklahoma and New Mexico together were a territory. (The word "territory" here reminds the author of the

Broadway show, "The Music Man" which had a song about "The Territory.")

The Harris Lion Bank went from a square base to an oval base because it was easier to cast and therefore cheaper. (Five pieces to three pieces.)

We made the trumpeting elephant bank for South America and the bull bank for Venezuela (tough banks to find). (They also made building banks for Mexico and Puerto Rico.)

Pig banks were big sellers for the company. We made lots of different pigs and a great number of each.

The Hopalong Cassidy bank was the first plastic banks we made. It was done in the 1940s. A badge came with the bank and there was a "Hoppy Club."

The cars were our biggest and best sellers representing over half of our sales for the company through the years.

Antiquo was a division in Banthrico that promoted the gift shop trade, especially in the 1980s. Bookend banks with music boxes were developed and we sold them directly to gift shops.

Banthrico was on Des Plaines Street when I joined the company in 1961.

The city of Chicago took our Des Plaines Street location property in 1970 and built the famous Claes Oldenburg Bat Column sculpture and the Social Security Building near it. We moved to Roosevelt Road in 1970 due to the Chicago property buyout.

Collecting

Banthrico banks can be found at flea markets, antique shows, malls, yard and garage sales, estate sales, antique and toy auctions, from private collectors and at museum sales. The latest source of Banthrico banks is on the Internet. There are several different sites that can be utilized. The easier to find Banthrico banks (A and B) are usually found at the flea markets, yard and garage sales, and antique malls. The more difficult Banthrico banks (C and D) are usually found in toy and antique shows and estate sales. The extremely difficult Banthrico banks (E - F) are usually found at toy and antique auctions, from private collections, and museum sales. The Internet is so new that all ratings can be found on it. There are, of course, exceptions to the above bank location areas and that is part of the fun of searching for Banthrico banks. You might find an "E" or "F" bank at a garage or yard sale on a card table or laying on a quilt in someone's front yard. By the same token, you may attend a toy show and find all kinds of Banthrico banks, especially those in the transportation chapter. If you attend a toy auction, you can expect to find Banthrico banks that are usually in great condition, much more costly, and lots of competition for them. These are the tough Banthrico banks you hope to add to your collection. At the antique and toy auctions and museum sales, you are not only bidding against the collectors sitting in the room (maybe 50 to 100), but there may be 100 to 300 other collectors that have submitted absentee bids for different Banthrico banks. There may even be someone on the phone bidding on the bank that you are bidding on. The Internet may have several thousand collectors looking at the items and who knows how many may be bidding on a particular item. All types of collectors besides Banthrico bank collectors are interested in the various Banthrico banks. The person bidding at an auction or on the Internet may be a building, car, financial institution, bust, animal, college, fictional character, registering or statue collector (and lots of other possibilities) that you are competing with for the Banthrico bank. Some collectors have picked up some great Banthrico bank bargains on the Internet and other advanced collectors have found that difficult "D" or "E" bank. A collector friend recently found the submarine Banthrico bank (an "E") on the Internet. He paid $240 for it and was delighted to add it to his collection. Another Banthrico collector has added some 150 Banthrico banks to his collection on the Internet (e-Bay) during the last year.

Auctions

There are several toy auction companies in the United States that are available for the Banthrico collector. The Henry/Peirce Bank Auctioneers in Chicago, Illinois, at 708/798-7508 have several bank auctions each year and issue a catalog for each sale. They usually have numerous Banthricos. Bill Bertoia Auctions of Vineland, New Jersey, also have several toy auctions (which includes banks) each year and issues a catalog for each sale. His banks are usually cast iron and tin, but Banthricos may be included in the sale. He can be reached at 609/692-1881. Gordon Steimle conducts a bank mail auction two or three times yearly and it has numerous Banthrico banks in it. His phone number is 414/454-0950. The Henry/Peirce and Bertoia catalogs usually cost $35-$40 each with several hundred bank and toy pictures in color with information on each item. Steimle's catalog is much smaller with some 100 banks done in black and white and usually free upon request. The catalog banks are illustrated with information on each. Bank catalogs with prices realized (provided after the sale as part of cost) become an important tool of the Banthrico collector. The catalogs can be filed and used for comparing picture, price, condition, date, and measurements of a given Banthrico bank.

Clubs

The best way to become knowledgeable of Banthrico banks and locate them in large numbers is to join a bank club. There is a "bank communication network" between several bank clubs and several hundred bank club members. A collector will learn more in one bank meeting or convention than he will alone in six months. When I joined the SBCCA in May, 1979, I had been collecting banks for several years and toys even longer. I was amazed how much I learned about banks so quickly and how my collection increased with some great buys from other SBCCA members. The Still Bank Collectors Club of America (SBCCA) is the premier national club of some 500 members and many of them collect Banthricos. The members collect all kinds of banks — cast iron, tin, lead, white metal (Banthrico), glass, wood, aluminum, vinyl, plastic, chalk, pottery, composition, cardboard, and combinations of the above. They have an annual convention (in a different city each year), a slate of officers, and publish several yearly newsletters and *The Penny Bank Post*, a thrice yearly periodical. At the annual 3 day convention, the member collectors display all types of banks in their hotel rooms to include Banthricos. The collectors have a great time together buying, bartering, trading, and selling banks. The annual convention also includes a free bank auction for members and Banthricos are always among the banks available. Bank collectors enjoy their food and fellowship and the convention includes a buffet and a seated dinner. The agenda includes educational programs presented on the different type banks. The club was founded in 1966. I am proud to be an active member and a former director of the SBCCA. Further information can be obtained by calling Larry Egelhoff, Membership Chairman, at 317/846-7228.

There are several local bank clubs that the beginning or advanced Banthrico bank collector may contact. There is the Gateway Penny Bank Club of the Greater St. Louis, Missouri, area which includes members from several Midwestern states. They meet every other month in the homes of the members. The members buy, sell, trade and have a great time "talking banks." Several Banthrico collectors to include the author are members of the club. A business meeting, "show and tell" session, and refreshments are included in the agenda. There are some thirty-five members. Further information can be obtained by calling Jim Redwine, President, at 314/741-6940. Another regional bank club is the Mid-Atlantic Penny Bank Club which meets in different cities in that area annually. They also display their banks in the various hotel rooms and buy, sell, and trade. Their annual meeting usually lasts two days. Their membership is about 150. For further information, contact Brian Cleary at 315/826-3610.

The Souvenir Building Collectors Society (SBCS) is a club that specializes in collecting souvenir buildings and many of their members collect Banthrico building banks. Their membership is around 35C and the organization is about four years old. They have an annual convention in cities around the country and also buy, sell, and trade in the hotel rooms. They have an auction, sightseeing trips, a raffle and usually eat at a local famous restaurant. The author liked the club so much that he recently joined. To receive further information on the group, contact Dixie Trainer, Editor of the *Souvenir Building Collector*, the SBCS Journal, at 804/361-1583.

Once you are a member of a bank club, especially the SBCCA, you are in a better position tc know private collectors and museum collections tha have banks available or will be put on the auctior block for sale. Many advanced collectors only trade banks and rarely sell them. When I first joined the SBCCA, I met Ralph Dye, who is one of the premie bank collectors in the country and I asked if he hac any banks to sell. His reply was simple. "I do no usually sell banks. I trade banks." That is also true of Andy Anderson who is "Mr. Banthrico." He prefers to trade Banthrico banks and his method seems to work. He has some 500 different Banthricos As a Banthrico bank collector, you need to be ir a bank club to have these kinds of contacts Many bank club collectors do sell their banks which allows a person to develop a collectio and secure some banks for upgrading and trading.

Reading the Picture Captions

The Banthrico banks have been cataloged int major design areas to assist the reader in locatin them. Each bank has been photographed separately unless there is a variation and a comparison is neede to show the difference. In Chapter 12, the ban boxes are shown with the related bank. Banks are listed alphabetically in each design area.

A majority of the banks were made with white metal. When Banthrico issued banks in other materials, such as lead, steel, plastic, composition, paper, etc., that will be indicated in the text of each chapter with the picture.

The information with each picture caption includes:

Number - each bank is numbered consecutively

Name and Location - of each bank. In some cases, the name of the bank may have variations. Some call the "ram" animal bank a "mountain goat" bank, and so forth. Keep this in mind as you use the book, and occasionally look for another possible name. Trophy banks (Chapter 7) are listed alphabetically by objects, not by the sponsoring organization. The building banks (Chapter 11) are listed alphabetically by the organization. Each chapter provides text on how to locate the banks in that chapter. The location of some banks will be given, if available, especially the architectural banks in Chapter 11. The locations of the colleges and universities also are provided. When the location is placed in parenthesis, the information is not on the bank and was secured from other sources.

Height - the height of the bank is provided (in inches, to the nearest eighth) for all banks except those in the transportation area. Banthrico made only one size of each bank, with only a few exceptions. White metal banks vary slightly (1/16 - 1/8 of an inch) from one batch or issue to another, due to manufacturing variations.

Length - the length of the bank will be provided only for transportation banks, unless indicated otherwise.

Trap (T) - The kind of trap in the bank will be indicated by a "T" and a number (example T4). Some 43 different traps have been identified on Banthrico banks. A separate chapter on traps with pictures and information will be located in the back of the book where the reader can see and study the type of trap that goes with a particular bank. Please note that Banthrico may have used several different traps on the same bank. If a bank was issued several times, each issue could have a different trap.

Generic or Specific - This will be indicated for each bank. If listed as generic (G), that indicates that numerous organizations used the bank to advertise their company. If listed as specific (S), that indicates that only one organization used (and usually owned the mold) the bank to advertise their company.

Wheel type - the wheeled vehicles in the transportation chapter will have a wheel type number with the picture. They are as follows:

1 - moveable rubber tire wheels, 1930s-1950s by National Products and Banthrico

2 - non-moveable WM wheels, 1950s & 1960s by Banthrico

3 - moveable spoked WM wheels, 1960s & 1970s antique vehicles by Banthrico

4 - moveable solid WM wheels, 1970s & 1980s modern cars by Banthrico

5 - moveable plastic wheels, currently by Valley Casting Co.

6 - moveable wooden wheels, earlier by Toystalgia Company

Year - The year or time frame the bank was introduced in the Banthrico line will be given. A "C" for circa (around or approximately) will be used when a definite date is not known, but an approximate time frame is known. The year an automobile or truck was manufactured by a company will be given, but that date will be different from the date the bank was introduced by Banthrico with the exception of promotional autobanks. The autobanks were issued the same year as the full size vehicles they portrayed.

Rating - The bank's rarity on a scale from A (the easiest to find) to F (the most difficult to find) is further explained earlier in this chapter.

Parentheses

The information in parenthesis has been added by the author from other research sources and is not on the bank. Several factors were used in arriving at the date of the bank, but this has been difficult because of the lack of Banthrico records. That is why you see numerous circas. If you have any information that would increase our knowledge of these banks, please let us know.

Chapter 2
People

U. S. Presidents

Of the sixty-one people banks known to have been made by Banthrico, eight are U.S. presidents and one is a presidential candidate. The presidents are D.D. Eisenhower, A. Hamilton, T. Jefferson, J.F. Kennedy, A. Lincoln, F.D. Roosevelt, T. Roosevelt, and G. Washington. Republican Thomas E. Dewey, Governor of New York, ran against Harry S. Truman in 1948 and lost the election. The Dewey bank is rarer than the Banthrico presidential banks. Several of the presidential bust banks have the signature of the president on the front of the bank.

Famous People

Nine Native Americans include three specific chiefs, named Anacostia, Pontiac, and Scioto; two tribes— Dowagiac and Shawmut; a woman; a child; and a rare seated Indian boy that is painted. Four explorers are among the people banks: LaSalle, a frontiersman at Fort Vancouver; Davy Crockett; and an astronaut. Great Americans such as Ben Franklin, Albert Pike, Will Rogers, and Robert F. Kennedy are among the group. Outstanding military leaders are not overlooked in the bust banks. They include Eisenhower, Lee, and MacArthur. Artists such as Charles Russell and Grant Wood, cowboy movie star Hopalong Cassidy, and inventor Thomas A. Edison, are among the bust banks.

Great American heroes such as The Minute Man and the Spirit of '76 were added in 1976 in Banthrico's Bicentennial Series. Three banks, "Tempo and Ed Felber," "Whizzo the Clown," and "Father Knickerbocker" need further research to fully identify them. Do you have information on these three banks you would share? The remainder of the banks are generic people such as the mariner and seaman, clowns, graduates, Hawaiian, Oriental Man, prospector, stock market investor and William Teller.

Abraham Lincoln leads the list in bank variations with five. Ben Franklin comes in second with four variations and George Washington with three variations as well as the Seaman at Wheel bank and the mechanical "Sav-O-Clown." Dwight Eisenhower and Douglas MacArthur both have two variations. So, the next time you are "bank hunting," take note of the variations. You may not have the Lincoln, Franklin, Washington, or others that you thought you have.

There are 23 rather difficult to find, "D," banks in this chapter. Five "E" banks are even more difficult to locate: Chief Pontiac, Dowagiac (Indian), Seated Indian Boy, Robert F. Kennedy, and Will Rogers. One rare "F" bank is Thomas E. Dewey.

When you examine Chapter 15 (Advertising Flyers and Letters), you will notice several composition and vinyl people banks that are not included in this chapter because none were found to photograph. I do not think these are difficult banks to find, but collectors do not realize they are Banthrico banks and are not collecting them. Also, some collectors do not collect composition or vinyl banks. The advertising flyers with the people shown also portray animals (#1034, 1035), insects, birds, etc. The flyers entitled "Custom Designed Replicas" (#1038) and "Half Pints" (#1036) were issued in the late 1970s and 1980s and list ceramic banks that could have been designed, but possibly not produced. The author does not know of an existing ceramic Banthrico bank.

1. Chief Anacostia
5", T4, S, c. 1955, D
Metal Inf. Plate on Back
Perpetual Bldg. Assn.
Washington D. C.

2. "Annie" Costia
4 3/4", T4, S, c. 1956, D
AFSLA, Anacostia Fed S&L - Washington, D. C.
Metal plate on back

3. "Andy" Costia
5 3/8", T4, S, c. 1957, D
AFSLA
Metal plate on back
Indian Papoose

4. Astronaut on Moon
3 ¼" (w/flag), T6, G, c. 1968, C
Slot in Back of Moon
Chippewa Trust Co. - St. Louis, MO
Bookend bank

5. Captain Marine
8 3/8", T12, G, c. 1975, C
Buoy on Back – Vinyl

6. Chief Pontiac
6 ½", T4, S, c. 1958, E
Chief Pontiac Fed Cr. Union
Pontiac, Mi.

7. Chief Scioto
", T4, G, c. 1960, D
Scoito Sav. Assn.
Columbus, Oh.

8. Clown, Blue Pl. Mechanical
7 5/8", T15, G, c. 1942, B
Called "Sav-O-Clown"

9. Clown, Multicolor Pl.
7 5/8", T15, G, c. 1944, C
Called "Sav-O-Clown"
Elliott State Bank
Jacksonville, Il.

10. Clown, Red Pl. Mechanical
7 5/8", T15, G, c. 1942, B
Came in gold also

11. Davy Crockett
5", T6, G, c. 1948, D
Bank of Brandywine
Brandywine, Md.

12. Thomas E. Dewey
5 1/8", T4, S, c. 1948, F
Governor of New York

13. Dowagiac (Indian)
6 3/8", T4, S, c. 1956, E
First Nat. Bk. Dowagiac, Mi.

14. Thomas A. Edison (Sig)
5 1/8", T4, G, c. 1958, C
"McGraw - Edison"

15. Eisenhower (General)
5 ½", T3, G, c. 1946, D
Slot in Back

16. Eisenhower (President. Sig)
5", T1, G, c. 1953, C
Metal Info Plate on Back
Mutual Fed S&L Assn.
Washington, D. C.

17. Father Knickerbocker
5 1/8 T4, S, c. 1962, D
Slot Also in Back of Hat
Knickerbocker Fed. Sav. , New York
N. Y. Dept. Store Logo?

18. Ben J. Franklin (Sig)
5 ¼", T4, G, c. 1940, D
American Nat Bk. of Granite City, (Il)

19. Ben J. Franklin (Sig)
5 ¼", T5, S, c. 1950, B
'The Franklin"

20. Ben J. Franklin (Sig)
5 ½", T6, G, c. 1956, C
Note Head Tilted Down
"To remove money loosen screws"
Bottom plate

21. Ben Franklin Savings
7", T12, G, c. 1978, C
Composition
"Made in Japan -
Made for Banthrico - Chicago"

22. Frontiersman with Rifle
4 ½", T4, S, c. 1972, D
Slot in Top – "Fort Vancouver"
First Fed Sav. -Vancouver, WA.

23. Graduates
5 ¼", T4, G, c. 1965, D
"Second Federal Savings"

24. Alexander Hamilton
5 ¼", T8, G, c. 1970, C
Hamilton Nat. Bk. of Knoxville, Tn.

25. Hawaiian Man with Helmet
5 ½", T4, S, c. 1966, D
Slot in Top Also
Hawaii State Sav. & Ln - Honolulu, (Hawaii)
"Ouku" on helmet

26. Hopalong Cassidy, Pl.
4 ¼", T35, G, c. 1950, D
Note Steerhead on Kerchief
"Bar - Twenty Associates"

27. Indian Chief
5", T8, G, c. 1962, B
(Native American)
Belleville (Il) Nat Sav. Bk.

28. Seated Indian Boy
6 ¼", T1, G, c. 1955, E
Unusual Painted WM Bank
Gramatan Nat. Bk & Tr. Co
Bronxville, New York

29. Th. Jefferson (Sig)
5 ¼", T4, G, c. 1952, C
First Nat. Bk.
Bowling Green, Oh

30. Robert F. Kennedy
7 3/8", T14, S, c. 1971, E
1925-1968
" c Handiguild Inc. 1971"
Converted bank

31. John F. Kennedy
5 1/8", T5, G, 1963, B
1917-1963
"Brunettin", (sculptor) on side of some banks
Security Tr. Co. - St. Louis, Mo.

32. La Salle
5 3/8", T4, S, c. 1968, C
La Salle Nat. Bk.
Chicago, Il.

33. R. E. Lee (Sig)
5 3/8", T4, G, c. 1955, C
Lee County Bk.
Fort Meyers, Fl.

34. Abraham Lincoln (Sig)
5 7/8", T1, G, c. 1940, D
Lincoln National Life Ins. Co.

35. Abraham Lincoln (Sig)
4 ¾", T2, G, c. 1946, D
Note Base Variation

36. Abraham Lincoln (Sig)
5 1/8", T1, G, c. 1951, C
Also T3
Archer-Hoyne Fed S&L Assn
"Savings insured to 10,000"

37. A. Lincoln (Sig)
5 1/8", T1, T4, G, c. 1962, B
Metal Name Plate Variation/Back
Gravois Bank (St. Louis, Mo)

38. A. Lincoln (Sig)
5 1/8", T8, T9, G, c. 1974, A
Newer Raised Name Area/Back
First Nat Bk. of Highland (Il)

39. Douglas MacArthur
5 ¾", T3, G, c. 1946, D
Slot in Back – Tall Base
Harpster Bk, Unicorp -
Harpster, Oh.

41. Man on Horse
5 ¾", T6, S, c. 1975, D
"Tempo and Ed Felber"
Who was Ed Felber and Tempo? Famous Royal Stallion Show?

0. MacArthur
5 ½", T3, G, c. 1946, C
lot in Back – Short Base
Jnited Ins. Co.
Chicago, Il

42. The Minute Man
5 ¾", T8, G, c. 1976, A
Bell and Flags on Back

3. Oriental Man with Mallet
5 7/8", T4, G, c. 1971, D
Oriental Writing on Sack
Daikoku, the God of Wealth - Japanese God of Fortune
Hawaii State Svg & Ln - Honolulu

44. Albert Pike
6 1/8", T4, S, 1959, D
1809-1891 – 150 Years
"Sesquicentennial of His Birth" (1959)
"Centennial as Sovereign Grand Commander"

45. Prospector
6 ¼", T4, S, c. 1972, C
The Bank of California

46. Will Rogers (Sig)
5 7/8", T1, G, c. 1948, E
Metal Info Plate on Back
Also done by Rehberger Co. for Preferred Bank Service Co.
Rogers County Bank (OK)

47. F. D. Roosevelt
4 7/8", T20, G, c. 1938, D
Western & Southern Life Ins. Co - Cincinnati, OH.

48. Theodore Roosevelt (Sig)
5 ¼", T4, G, c. 1958, D

49. C. M. Russell (Sig)
6 3/8", T8, G, 1962, C
Note Steer Skull with Sig
"C. Dennis - 1962"
"© Bancorporation of Montana
Helena, Mt."

50. Seaman at Wheel
6 ½", T4, S, c. 1955, D
Note Ship in Wheel – 1st
Marine Bank & Tr. Co - Tampa, Fl - Founded 1914

51. Seaman at Wheel
6 ½", T4, S, c. 1955, C
Note Ship in Wheel – 2nd
Banco Del Caribe (Carribean?)

52. Seaman at Wheel
6 ½", T4, S, c. 1955, C
Note Ship in Wheel – 3rd
Marine Midland Bank

53. Shawmut (Indian)
6 5/8", T4, T10, S, c. 1972, C
"Shawmut Banks" (Boston)

54. The Spirit of '76
6", T9, G, 1976, A
Bicentennial Year
Mercantile Bank - St. Louis, Mo

55. Stock Market Investor
6 ½", T12, G, c. 1980s, C
Paper "Banthrico" Sticker/Bottom
Composition

56. G. Washington (Sig)
5 1/8", T1, G, c. 1947, C
Metal Info Plate on Back
Lacy's Inc. - Washington, D.C.

57. G. Washington (Sig)
5 1/8", T4, G, c. 1954, B
Newer Raised Name Area
First Nat. Bank - New Bedford, MA.

58. G. Washington (Sig)
4 5/8", T23, G, c. 1940, D
Small Early Version

59. Whizzo the Clown
6 5/8", T4, S, c. 1966, D
a/k/a Whizzer
Home Sav Assn of Kansas City (Mo)
Note his big feet. Who was Whizzo?

60. William Teller
6 3/4", T12, G, c. 1980s, C
Paper "Banthrico" Sticker/Bottom
Composition
Bank "teller" in name

61. Grant Wood
5 1/8", T4, G, c. 1948, D
1891-1942
Iowa Artist who painted *American Gothic* (Standing farmer with pitchfork and wife)
Citizens Sav. Bank - Anamosa, Ia.

Chapter 3
College and University Mascots and Sports

The college and university mascots and sport anks are some of the most interesting, creative, nd unusual banks that Banthrico produced. The nthusiasm today for sports and alma maters makes nese banks highly collectible. The majority are dif-cult to find. Banthrico made these white metal anks from the late 1940s into the 1980s. In the 970s and 1980s they changed to making them 1 composition and vinyl, like the Massillon Tigers nd Misha. The majority of the white metal mascot anks were introduced in the 1950s and 1960s, vith some in the 1970s. These banks are usually ound in fine or better condition because often they vere nostalgic, were handled very little, and were packed away" for safe keeping.

The college and university mascot banks com-rise the largest group in this category, but there re also two high schools (Massillon Tigers and Crown Point Bulldogs) and two unknown mascots Columbia Bulldogs and Conroe Tigers). One base-all team is among the mascots, the St. Louis Car-inals of the National League, with a great baseball istory. It is a hard bank to find. It was issued by nly one bank. There are some beautiful and de-ailed banks in this group.

One football player, Sammy Baugh, was also resident of an insurance company. An NFL foot-all, three college bowl footballs, and the 1980 ummer Olympics with Misha are among the sport anks. A Major League baseball player is included vhich was a Campbell Soup Co. promotion. The USL" Bulldogs location is an "educated guess" by some of the advanced Banthrico collectors; we may be wrong, let us know if we are. The unusual woman bowler trophy bank is probably a brass pattern bank. The base is brass, the woman is white metal, and the trap, identified as "Banthrico, Inc.," is white metal in a steel bottom plate.

The state of Iowa has the most banks represented with six, Texas has five, Missouri, Alabama, Louisiana and Ohio have three each, and several states have two. Two college powerhouses, Notre Dame and Purdue, are both located in the Hoosier state. Banthrico did not produce a mascot bank from Illinois. There are 31 animals and birds, 10 people, an insect, and a weather condition (cyclone) represented in the mascot and sport banks.

There are twenty-four "D" banks and eleven "E" banks in this category. There is one "F" bank, the woman bowler trophy.

63. Alabama Crimson Tide
(University of AL, Tuscaloosa)
6 1/8", T4, G, c. 1950s, E
Slot in back

62. Air Force Academy (Falcons)
(Colorado Springs, CO)
6", T4, G, c. 1970s, D

64. Arkansas Razorbacks
(University of AR, Fayetteville)
4 3/4", T13, G, 1970s, C
"Woo-oo Pig Sooey"
Pulaski Fed Sav (PFS) -
Little Rock, (AR)

65. Auburn Tigers
(Auburn University, Auburn, AL)
6 1/8", T4, G, c. 1960s, C
Home Fed. Sav - Columbus, Ga.

66. Auburn Tigers
(Auburn University, Auburn, AL)
4 1/8", T13, G, 1950s, C
Slot in trap and behind tiger

67. Baseball Player on Base
5 7/8", T13, G, c. 1950s, D
Major League Baseball
Campbell Soup Co.offer

68. Sammy Baugh, Quarterback
5 1/8", T4, S, c. 1950s, D
NFL Hall of Fame
Southwestern Fidelity Life Ins. Co. -
Houston, Tx.

69. Baylor Bears
(Baylor University, Waco, TX)
6 3/8", T6, G, c. 1970s, D
Lake Air Nat. Bk of Waco, Tx.

70. Columbus Bulldogs
(Columbus, OH)
6", T6, G, c. 1960s, D

1. Conroe Tigers
Conroe, TX)
3/8", T6, G, c. 1960s, D

72. Crown Point Bulldogs
(IN High School)
6", T6, G, c. 1960s, D

73. Georgia Bulldogs
(University of GA, Athens)
6", T4, G, c. 1950s, D
Ashburn Bk - Ashburn, Ga.

4. Georgia Tech Yellow Jackets
GA Inst. Of Tech, Atlanta)
1/8", T4, G, c. 1960s, E
ank of Toccoa - Toccoa, Ga.

75. Iowa (Hawkeyes)
(University of IA, Iowa City)
5 ½", T1, G, c. 1948, D
Metal name plate "Iowa" on helmet - "24" on jersey
Slot in back also - Iowa State Bk & Tr. Co. - Iowa City, Ia

76. Iowa State (Cyclones)
(IA State University, Ames)
6 ¾", T6, G, c. 1960s, D
"I" on jersey

Left: 77. Iowa (Hawkeyes)
(University of IA, Iowa City)
5 ½", T6, S, c. 1950s, C
Slot in bottom and in back of hawk
"Officially licensed - Univ. of Ia."

Right: 78. Rose Bowl Football
4 ½", T10, S, 1981, E
Iowa Hawkeyes – First Ed. – 1981
"1981 Big 10 Co-Champions"
"8 - 4" (Record)

79. Peach Bowl Football
4 ½", T10, S, 1982, E
Iowa Hawkeyes – 2nd Ed. – 1982
"8 - 4" (Record)

80. Gator Bowl Football
4 ½", T10, S, 1983, E
Iowa Hawkeyes – 3rd Ed. – 1983
"9 - 3" (Record)

Left: 81. Lenoir Rhyne College (Tigers)
(Hickory, NC)
6 ½", T6, G, c. 1960s, E

Right: 82. Louisiana State University
(Tigers – Baton Rouge, LA)
6 ¼", T4, G, c. 1960s, D
Bank of Louisiana

33. L. S. U. Tigers
LA State University, Baton Rouge)
1 1/8", T3, T13, G, c. 1950s, C
Slot in trap and behind tiger

84. University of Maine (Bears)
(Farmington, ME)
6 ½", T4, G, c. 1960s, D

85. Massillon (High School)
(Tigers – Massillon, OH)
6 1/8", T6, G, c. 1960s, E
Massillon Sav & Ln Co
(Massillon, Oh.)

36. Massillon (High School)
Tigers – Massillon, OH)
7/8", T12, G, c. 1980s, C
amous high school football
eam once coached by legendary
NFL Coach Paul Brown - Stark
County, Ohio- Home of NFL Hall
f Fame-
Composition

87. Michigan State (Trojans)
(MI State University – East Lansing)
7", T4, G, c. 1950s, D
Michigan Nat. Bk. - Port Huron, MI

88. Michigan Wolverines
(University of MI, Ann Arbor)
5 3/8", T4, G, c. 1950s, D
Founded in "1817"
Birmingham Fed Sav. (MI)

Left: 89. Minnesota Golden Gophers
(University of MN, Minnpls-St. Paul)
6", T4, G, c. 1950s, D
University Nat Bk. - Minneapolis (MN.)

Right: 90. Missouri Tigers
(University of MO, Columbia)
6 1/8", T8, G, c. 1960s, D
Kansas City Bk & Tr. Co -
(Kansas City, Mo.)

Left: 91. Missouri Tigers
(University of MO, Columbia)
4 1/8", T13, G, c. 1950s, C
Columbia Sav. Bk - Columbia, Mo.
Slot in trap and behind tiger

Right: 92. Nebraska Cornhuskers
(University of NE, Lincoln)
6 3/8", T4, G, c. 1970s, D
First Nat. Bk. - Grand Island, Ne.

Left: 93. NFL Football – Vinyl
4 1/8", T38, G, 1976, B
National Football League
AFC and NFC paper team helmet stickers

Right: 94. Notre Dame Fighting Irish
(Notre Dame, IN)
6", T4, G, c. 1950s, D
Cloverleaf on hat

95. Oklahoma State (University)
Cowboys – Stillwater, OK)
5", T4, G, c. 1960s, C

96. Oklahoma (University)
(Indians – Norman, OK)
6 1/8", T4, G, c. 1960s, C
Liberty Nat. Bk & Tr. Co of
Oklahoma City, (OK)

97. Oregon State University
(Beavers – Corvallis, OR)
6", T4, G, c. 1950s, E

98. University of Oregon
Ducks – Eugene, OR)
1/8", T4, G, c. 1950s, D
O" on jersey
Citizen Bank - Corvallis, Or.

99 Purdue Boilermakers
(Purdue University – Lafayette, IN)
6 ¾", T4, G, c. 1950s, D
Man with hammer and football standing on a boiler
The Bank - Indiana National

100. Rice University (Owls)
(Houston, TX)
5 ½", T4, G, 1962, E
"The Academic Seal of Rice University - 1912-1962"

Top left: 101. St. Louis Cardinals
(St. Louis, MO)
6 ½", T4, T6, S, c. 1948, E
Gravois Bank - St. Louis, Mo

Top center: 102. UT – Texas (Longhorns)
(University of TX, Austin)
5 ¾", T4, G, c. 1950s, E
"Bevo" on helmet, "00" on jersey

Top right: 103. "USL" (Bulldogs – Lafayette)
(University of Southwestern LA)
6", T4, G, c. 1960s, D

Center left: 104. Wisconsin Badgers
(University of WI, Madison)
6 ¼", T10, G, c. 1950s, D
Badger Savings

Center: 105. Bucky Badger – Comp.
(University of WI, Oshkosh)
7", T12, G, c. 1980s, C
"Made for Banthrico - Chicago"

Center right: : 106. Woman Bowler Trophy
4 7/8", T5, G, c. 1950s, F
Brass Pattern Bank

Bottom center: 107. Wyoming Cowboys
(University of WY, Laramie)
6", T4, G, c. 1960s, C
American Nat. Bk. Cheyenne, WY

Bottom right: 108. Misha, Olympic Bear
5 ¾", T12, G, 1980, B
Official Mascot/Summer Olympics
" © Image Factory Sports Inc."
Moveable head and arm
"Prod. of Hong Kong" - Vinyl

Chapter 4
Fictional Characters

All the fictional character banks shown here are plastic or vinyl, except four of the Santa banks. The medium and large sleeping Santas, plus the Santa on Roof are white metal. The Santa in Large Chair is chalk (Banthrico called it plaster).

The oldest bank in this section is probably the Jack-in-the-Bank (Box) which was made in the 1950s. It is fragile and difficult to find all intact and working, since it is a mechanical that speaks (but is difficult to understand). The Sleeping Santa in Large Chair was manufactured in 1954 and a few years following, but the white metal sleeping Santas (medium and large) followed the next two years and eventually replaced the chalk Santa. The two white metal sleeping Santas were big sellers at Christmas time. It was not unusual for the Santas to each outsell any of the other single banks for the entire year! The Santa on Roof came along in four years and soon replaced the white metal sleeping Santas. It was the big Christmas promotion during the 1960s. You will note in one of the pictures in the book that this Santa has a clear plastic bag of miniature toys. The bag of toys usually gets lost from the Santa (after all, the banks were made for children, too) and it's rare to find one. Thanks to Marty Gutirrez, we have both the Santa and his bag of toys. The small sleeping vinyl Santa appeared in the 1970s and became the Christmas attraction for Banthrico. I feel certain that the small vinyl Santa did not sell as well as the white metal Santas for Banthrico. If you look closely at the two sleeping Santas (medium and large) you will notice some minor differences. The tilt of the head and the hat are two examples. These two sleeping white metal Santas in the Chair did so well that Banthrico kept that same bank mold idea when they changed over to vinyl. Oh! How we wish as bank collectors they had stayed with white metal in these and many other banks they made, but to compete and survive, they had to look to inexpensive materials and processes. And, they still did not survive.

Mechanical Banks

There are five mechanical banks in this chapter. Banthrico did not manufacture or market many mechanical banks. There are a few in the other chapters, but only one (Goose) (#1032) was made of white metal. All the other mechanicals are plastic or vinyl. There are some registering and calculating banks in steel and they will be covered in a later chapter. The three mechanical Walt Disney banks — Donald Duck, Goofy and Mickey Mouse — were all made by Paragon-Reis and distributed by Banthrico in the early 1980s. They are well done mechanical banks and the boxes they came in are quite colorful. One of the boxes is shown in the chapter covering the boxes. Bozo the Clown is always a favorite subject for bank makers and Banthrico was no exception. It is a colorful mechanical bank using a very familiar mechanical action. The Jack-in-the-Bank was discussed earlier. It has four slots on the top of the red box for 10 cents, 1 cent, 5 cents, 25 cents and any coin would activate Jack to moving and speaking.

Little LuLu has been a favorite for a number of generations. She spans the last five decades and joined the Banthrico line in the 1970s. Frosty the Snowman fits in well with the Santa Claus banks and is always a hit with the kids and the collectors. Fredrick the Fox (late book arrival) and the Grape Kid (what is his real name?) were recent additions to the Banthrico vinyl bank parade in the 1980s.

There are several fictional character banks in the Banthrico advertisement flyers (#1038) that are not pictured in the book. Maybe more of these banks will become available once bank collectors realize they are Banthrico and they start collecting them. There is the possibility that some of the fictional character banks shown in the flyers never made it into production since they were custom designed and had to be ordered by the customer and then made by Banthrico.

There are five "D" banks in the chapter. They are Bozo, Jack-In-the-Bank, Little LuLu, and Sleeping Santa-medium and large. There are no "E" or "F" banks in this group.

Let me close this chapter by predicting that this will become a big bank collecting specialty in the years ahead. It is already an area of emphasis for the author with several hundred fictional character banks that includes newspaper, radio, advertisement, movie, and TV characters. Many bank collectors have not collected any bank materials except metal. However, the plastic, vinyl, chalk and composition banks are colorful, historical, easily obtainable and usually inexpensive and should be considered. Banthrico fictional character banks will be a nice addition to this specialty area.

Left: 109. Bozo the Clown
7", T15, G, 1972, D
Larry Harmon Pictures Corp.
"Bozo" on sleeve - mechanical

Right: 110. Donald the Golfer
7 ¼", T19, G, 1981, B
"Walt Disney Productions"
"Paragon-Reiss - Copyright 1981"
mechanical

111. Frosty the Snowman
7 ¾", T12, G, c. 1970s, C
Vinyl
"Banthrico - Chicago" on Back

112. Goofy the Soccer Player
7 ½", T19, G, 1981, B
"Walt Disney Productions"
Button on side starts mechanical action

113. The Grape Kid
5 ½", T12, G, c. 1980s, C
"Banthrico – Chicago"
Vinyl - plastic stopper - "Niagara Plastics"

114. Jack-in-the-Bank
6", T19, G, c. 1950s, D
Plastic Mechanical
Four slots in top for 10, 1, 5, 25 cent coins
New Bedford and Acushnet Coop. Banks (MA)

115. Little LuLu
8", T12, G, c. 1970s, D
Vinyl – "Banthrico-Chicago"
"TM and © Western Publishing Co. Inc."

116. Mickey the Magician
7", T19, G, 1981, B
("Walt Disney Productions")
Vinyl mechanical

117. Santa on Roof
6 ½", T6, G, 1960, C
Rare Bag of Small Toys
First Granite City Nat. Bk., Granite City, Il

118. Sleeping Santa – Small
5 ¾", T12, G, c. 1970s, C
Vinyl

119. Sleeping Santa – Medium
6 1/8", T6, G, 1956, D
Slot in Back
First Nat Bk of Wood River, Il

120. Sleeping Santa - Large
6 ¾", T6, G, 1955, D
Slot in Back
Columbia Fed Sav & Ln Assn., Washington, D.C.

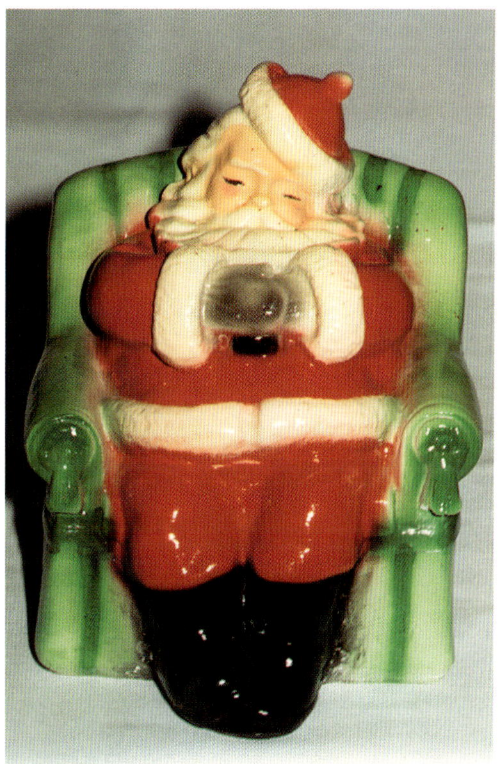

121. Sleeping Santa – Large Chair
7 1/8", T16, G, 1954, C
Chalk – Slot in Back
Chair comes in assorted colors.
Banthrico "plaster".

122. Fredrick the Fox
8", T12, G, c. 1980s, C
Composition
Bottom paper sticker - "Made for Banthrico - Chicago"

Chapter 5
Animals, Birds, and Fish

Living creatures of the land, air and water comprise this group, including forty-two animals, six birds, two fish, and one frog.

Animals

There are eight dog banks and eleven pig banks, which always seem to be favorites of bank makers and collectors. Several color variations exist among the pigs, which increases their representation.

Dogs

The dog banks include at least seven different breeds, and one cannot be certain about the Dog Waving bank. Most collectors do not realize that the Begging Dog (#135) is a Banthrico (Moore #408 - Forlorn Dog), nor did Andy and Susan Moore. So, collectors check your collections and your favorite bank hunting places! The Seated Scottish Terrier (#141) is a very tough bank to find. It is rated a "D," but it is close to an "E" bank. The Dachshund is a semi-mechanical bank that expands with the coins inserted (and some help from the collector pulling the head and tail in opposite directions). You can also see the box and bank together in the box chapter. The hound, poodle and Schnauzer banks are recent additions, with the seated cat, to the Banthrico inventory in the 1970s. The cat and mouse heads are really nice banks and Banthrico called them "Jewelry Coin Banks" because of the glass eyes, the brushed texture effect in the metal and the gold plate.

Banthrico made numerous pig banks, including a big composition pig bank with flower designs in the 1960s that is shown with the advertisement flyers. The pigs were done in many colors such as the Happy Colorful Pig, the Pig with Open Mouth and the Red Pig with Big Ears. The slot in the Pig with Open Mouth is a rather unique approach and makes it a very interesting bank to collect. The Top Hatted Pig is my favorite among the pigs. Banthrico called him "Sir Pig" and dressed him in a tuxedo in black, yellow or red. The white shirt bib in front pulls down to insert or remove the coins. The bank is a 1950s plastic product. The black pig was not popular (which surprises me) with the customers and was discontinued by Banthrico (which makes it harder to find than the yellow and red variations).

The ram, cougar, eagle and pelican (on posts) were made on the 1970s. They were all made from hand carved wood originals and they are gorgeous banks! One of the Banthrico flyers stated they represented "some of America's most majestic and renowned wildlife animals and birds that are fast disappearing from the American scene" (1970s). They are all very popular banks and highly collectible. They come in a beautiful copper color. Banthrico also made and marketed these as bookends as well as banks and if you examine the backside of the bank, it is flat for books. Many of these banks were sold directly to gift shops and other outlets and they had no advertisement on them. Many of them also included music boxes and the winding key was in the base trap.

The trumpeting elephant and the buffalo lying down are probably the two hardest banks to find in this group. The small buffalo, black Brahma bull, begging dog, seated Scottish terrier, fat pelican, Rath pig, two seated bears, and pike with metal nameplate are also difficult banks to locate. This group of banks well illustrates the rating of banks from "A" to "E." There is a balanced representation at every level. There are no "F" banks in this group.

Birds

Six bird banks were made, which comes as a surprise since there are so many birds available to choose from. The ducks and the pelicans lead with two each. The Fat Pelican bank was made for a bank in New Orleans, Louisiana, and a bank in New Jersey. It is an early bank and difficult to find, therefore rated "D." The eagle is becoming difficult to find and is rated "C," and it is one of the most beautiful banks that Banthrico made. It is also a popular bank with collectors since it is the national bird and seen on official symbols and documents.

Others

One group of banks that is shown in the advertisement flyers is the "warm fuzzies" (#1038). The banks in this group never became available from the bank collectors, nor did the author have them in his collection. They are flocked banks (thin vinyl with fuzz) and the majority of bank collectors I know do not collect them. I am including them here so the reader will know they are Banthrico banks and may choose to add them to their collection. There are four different sets in the flocked banks. The metal "Golden Egg Goose" mechanical bank (#1032) was issued by Banthrico in the late 1930s or early 1940s and is not included. No one seems to have it so it is shown only in the advertisement flyer. It appears to be about 6" high from a picture. When you feed the goose coins, his wings, tail and bill moves. It retailed for $1.00. Another bank not shown, yet we know it exists because of a Banthrico flyer, is the "Gulping Frog." A mechanical plastic frog that swallows coins. It came in cherry red and will be with the flyers. Another trio of banks not shown that aren't animals, birds, or fish, but they are living creatures so I will place them here. They are insect banks (#1039) "Caterpillar," "Lady Bug," and "Grasshopper." They came in assorted bright colors and were made of composition (papier-mâché per Banthrico). Peter Eisendrath (the last Banthrico president) told me a funny story about these banks. Banthrico had Japan to make these banks for them (as they did numerous other composition banks) and when they arrived by boat in San Francisco, California, the forklift truck dropped the total shipment in the Pacific Ocean. Needless to say, Japan had to make and deliver another bank shipment. These banks were made in the early 1980s and had a removable rubber plug in the bottom. They are shown in the flyers and appear to be very collectible.

There are several vinyl animals that were produced in the 1980s — Koala Bear, St. Bernard, Kangaroo, Raccoon, and Squirrels. These banks were well done, quite desirable, and extremely durable. You see Banthrico moving from metal to vinyl and trying to diversify and keep the company prosperous. They still made white metal banks in the 1980s, but it was no longer their "bread and butter" like it was in the 50s, 60s, and 70s. The company ended production in mid-1985 and closed their doors forever.

123. Koala Bear with Cub
6 ¾", T12, G, 1980s, B
Vinyl
"Banthrico - Chicago"

125. Seated Bear
4 ¾", T5, S, c. 1960s, D
"Thrifty McBear"
Nat. Bk of Alaska - Kodiak, AK.

124. Seated Bear
4 ¾", T4, S, c. 1950s, D
"Established 1887"
Calif. Sav & Ln. Co.

127. Buffalo Lying Down
4", T5, G, 1969, E
"J. Contway – 69" (Sculptor)

126. Large Buffalo on Base
6", T4, S, c. 1950s, C
Replica of 13 ton statue in facade of North American Life and Casualty Co. office in Minneapolis - Carved from Mn. granite by Sculptor John Daniels

128. Small Buffalo in Grass
3 ¾", T5, S, c. 1960s, D
" 1st National Bank of Greybull"

Left: 129. Brahma Bull on Base
5 ¼", T8, S, c. 1960s, C
"Banco de Los Llanos"
(Bank of the plains or flat lands.)
Brands on bull

Right: 130. Brahma Bull on Base
5 ¼", T8, S, c. 1960, D
Black Painted Variation

Left: 131. Bull on Ear of Corn
4 ¼", T8, S, c. 1970s, C
"Since 1910"

Right: 132. Cat Head
3 ¼", T3, G, c. 1960s, C
Green Glass Eyes
Slot in back of head -
Southern Commercial Svg Bk - St. Louis, MO

Left: 133. Seated Cat
5 ¾", T9, G, c. 1970s,

Right: 134. Cougar on C
4 7/8", T10, G, c. 1970
Music Box in Base

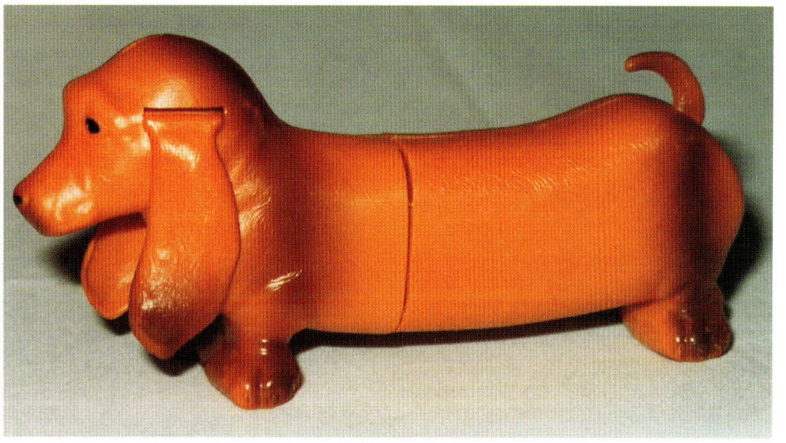

136. Dachshund Dog
3 ¾", T36, G, c. 1960s, C
Plastic Semi-Mechanical
"Banthrico - Chicago" under left ear. Slot in top of head.

135. Begging Dog
4 7/8", T3, G, c. 1950s, D
Slot in Top

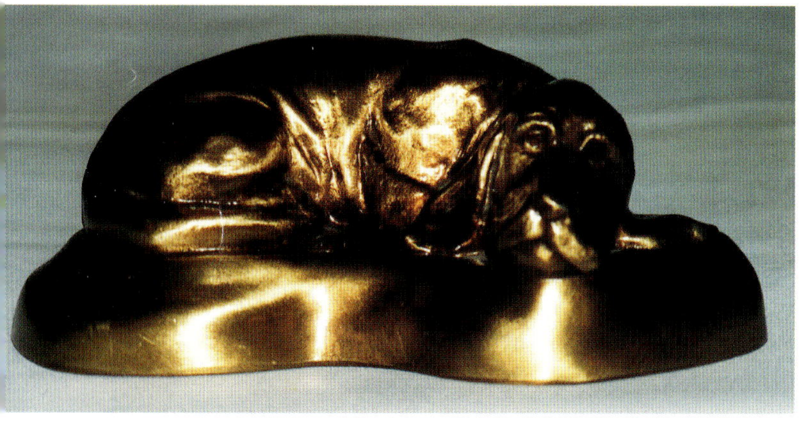

137. Hound Dog Lying Down
2 ¼", T9, G, c. 1970s, B

138. Poodle Lying Down
4 ¼", T9, G, c. 1970s, B

Left: 139. Seated St. Bernard
6 ½", T12, G, c. 1980s, B
Vinyl

Right: 140. Seated Schnauzer
6", T9, G, c. 1970s, B

Left: 141. Seated Scottish Terrier
5", T4, G, c. 1960s, D
"Thrifty Mac" with Bow
Tampa Fed S&L Assn. (Tampa, Fl)

Center: 142. Dog Waving
8 1/8", T12, G, c. 1980s, B
"Your Money Friend"
"Banthrico - Chicago" on Bottom

Above: 143. Baby Duck
4 ½", T10, G, c. 1980s, A

144. Mallard Duck
3 ¼", T10, G, c. 1980s, A
"L. Masten ©"

145. Eagle on Cliff
6 ½", T8, G, 1976, C
Americana Bicentennial
Smith County State Bk & Tr. Co. -
Smith Center, Ks

146. Trumpeting Elephant
6 1/8", T4, S, c. 1950, E
Also in Pewter Color – T10, C
Banco Del Carbide, C.A.
Su Banco De Confianza

147. Fat Frog
3 ¼", T12, G, c. 1980s, B
Vinyl – "Banthrico-Chicago"

148. Kangaroo with Baby
6 7/8", T12, G, c. 1980s, B
Vinyl – "Banthrico-Chicago"
Continental Bank

149. Harris Lion
6", T4, S, c. 1960s, C
Oval Base
Harris Trust and Savings Bank

50. Harris Lion
", T4, S, c. 1950s, C
quare Base

151. Monkey with Drum
4 ¼", T6, G, 1961, C
"©1961 - Unique Coin – Chicago"
Security Nat Bk of Battle Creek (MI)

152. Mouse Head
3 3/8", T3, G, c. 1960s, C
Red Glass Eyes
Southern Commercial & Sv. Bk. - St. Louis, Mo.
Jewelry Coin Bank

153. Small Owl
3 1/8", T12, G, c. 1980s, B
Vinyl – "Banthrico-Chicago"

154. Fat Pelican
6 ¼", T1, G, c. 1950s, D
Slot in Back
Pelican Homestead Assn.
New Orleans, La.

155. Pelican on Pier Posts
6", T9, G, c. 1970s, C
Music Box in Base
First Granite City Nat. Bk. - Granite City, Il.

156. Fillmore Pig
3 ½", T8, G, c. 1970s, B
Flower Design on Sides
"Fill me up with more coins"
Bohemian S&L Assn. - St. Louis, Mo.

157. Happy Colorful Pig – 1
3 ¼", T26, G, c. 1960s, B
Tan – Slot in Top
Central Missouri Trust Co. -
Jefferson City, Mo.

158. Happy Colorful Pig – 2
3 ¼", T26, G, c. 1960s, B
Coral – Slot in Top
Central Colorado Bk. - Colorado Springs, Co.

Left: 159. Happy Colorful Pig –
3
3 ¼", T26, G, c. 1960s, B
Red – Slot in Top
South Side Nat. Bk in St. Louis (Mo)
Banthrico "Blubber Boy"

Right: 160. Jolly Pig
3 3/8", T6, G, c. 1955, B
Made in Silver and Gold
Jefferson - Gravois Bank of St. Louis
Slot in top - Changed to bronze and brass finishes

eft: 161. Pig with Open
Aouth
7/8", T26, G, c. 1960s, B
qua – Slot in Mouth
anthrico "Money Grabber"

ight: 162. Pig with Open
Aouth
7/8", T26, G, c. 1960s, B
ellow – Slot in Mouth

Left: 163. Rath Pig
4 ¾", T10, S, 1983, D
"1933 – 1983"
"Rath Employees Credit Union"

Right: 164. Red Pig with Big Ears
3 ½", T3, G, c. 1940s, C
Slot in Top
Snyder County Tr. Co. - Selinsgrove, Pa

Left: 165. Short Oink Pig
3 ½", T6, G, c. 1970s, A
Flower Design on Sides/Ears

Right: 166. Top Hatted Pig
5 ½", T28, G, c. 1950s, C
Plastic
Almira S&L Assn. - Chicago, Il.
Banthrico "Sir Pig"

Left: 167. Pike Jumping in Water
6 7/8", T4, G, c. 1950s, D
Metal Name Plate
Woodruff State Bk - Woodruff, WI.
Slot in mouth

Right: 168. Pike Jumping in Water
6 7/8", T8, G, c. 1960s, C
Engraved Name Variation

169. Baby Raccoon
3 3/8", T12, G, c. 1980s, B
Vinyl – "Banthrico-Chicago"

170. Ram on Mountain
6 5/8", T6, G, c. 1970s, C
Music Box in Base
Tune "Oh What a Beautiful Morning"

171. Squirrels on Base
6 ¼", T12, G, c. 1980s, B
Slot in Stump Behind Squirrels
"The Squirrels Club - Banthrico - Chicago"
What does "F" stand for?

172. Thrifty Squirrel
6 7/8", T29, G, c. 1950s, C
Plastic – Mechanical
Mont Clare Svgs., Elmwood Park, Il.

173. Teddy Bear Seated
5 1/8", T10, G, c. 1980s, A

Chapter 6
Household Items and Food Products

Household items and food products are closely related design ideas concerning shelter and nourishment. Twenty-five household item banks and eight food product banks were selected as items that were found in a typical home between 1900 and 1960.

In the House

Early homes had wooden barrels for apples, pickles, and sauerkraut. A baby's book to record the child's experiences and growth, and a rocking horse for the children's play could be found in many homes. Items related to the mothers include a dressmaker dummy for measuring children's and adults' clothes, a sewing machine for making them, and a big kitchen stove where mom and grandma fixed all the wonderful meals.

The advent of the vacuum sweeper and broom led to the invention of the vacuum cleaner, and two model banks are illustrated. There is also a 1908 Speed Queen Washer.

The teenage girls in the home might especially be interested in the jewelry box, record player, tape recorder, and Bulova watch. Dad especially liked the "Snug Harbor" chair for relaxing and the suitcase and umbrella for traveling and working.

Food Items

There are three ears of corn and one of them is with a bull. The bull on the ear of corn bank is also in the animal section, since it relates to both sections. This bank represents both beef and corn food sources. There are three varieties of Idaho potato banks, and #206 is the more difficult one to locate. The bee provides honey, so the Beehive with Bee bank was placed with the food products.

The original carved coconut bank was made by Paul Gauguin, the famous French artist, on the island of Tahiti around 1891, and was owned by Philip Ward who lived in Chicago. Banthrico and Ward came up with the idea of making a WM bank from the coconut; it is probably the most unusual bank Banthrico made. In the shape of an egg, it has the face of a fish and carvings of Tahitian women and palm trees on the sides. It was especially promoted by Banthrico at art shows and galleries and is difficult to find today, so is rated a "D" bank.

There are several "D" and "E" banks in this chapter. The oak barrel, furniture cabinet, "Snug Harbor" chair, dress dummy, sewing machine, two vacuum cleaners, beehive, coconut and Idaho potato on end are all "D" banks. There are four banks that merit an "E" rating — the rocking horse, record player, tape recorder, and Bulova watch.

The rocking horse bank is extremely difficult to locate and may eventually be rated an "F" bank. The banks were issued by a small company in Winchendon, Massachusetts, that was known for its well made wooden toys, so the town became known as "Toy Town." The rocking horse bank was a natural for them and for Banthrico. The colorful flyer on the town is in the Chapter 15. The rocking horse is also the bank logo for the SBCCA.

The Pentron record player and tape recorder are likewise very hard to find and they are usually found in the lower condition levels (fair to good). The Bulova watch was first manufactured by National Products before Banthrico purchased the company in 1940. Banthrico continued this bank through the early 1940s and again after World War II, in the late 1940s. It is rarely seen and is rated "E."

Banks shown in Chapters 8, 13, and 15 are "miniature typewriters." They would have been put in this chapter, but were not available at the time. They came in gold and silver and were made by National Products and then by Banthrico, both located at 560 W. Lake Street in Chicago. They could be purchased for 25 cents each for quantities of 100 or more (30 cents each for smaller numbers) during the late 1930s and early 1940s. They are "C" and "E" banks.

175. Baby Book
4 ½", T10, S, c. 1970s, C
"Book of Thrift"
Safeway Employees Assn - SEA

174. Wooden Oak Barrel
5 ¾", T21, G, c. 1930s, D
Steel rings and ends
Hand Coopered Barrel - Original - Bankers Thrift Corp. Product
"Pioneer Barrel" - 1000 coin capacity
Lots of 500 - $1 each

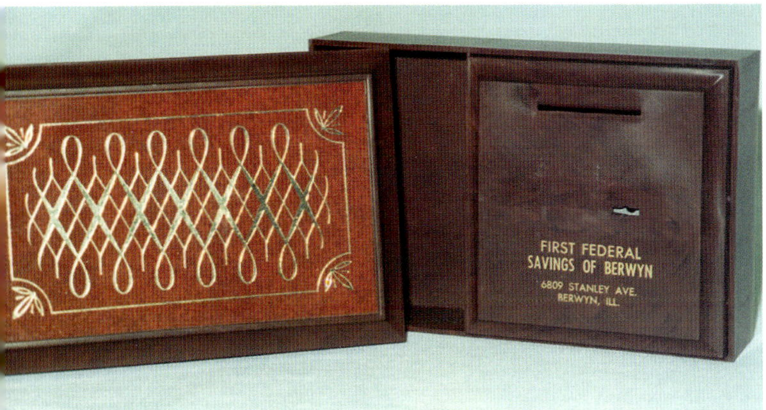

76. Jewelry Box
", T22, G, c. 1960s, C
ote Scrolled Lid
rst Fed. Svg. of Berwyn - Berwyn, Il.

177. Furniture Cabinet
5 5/8", T3, S, c. 1950s, D
"Union National, Inc."
"Fine Furniture -
Jamestown, N.Y."

78. "Snug Harbor" Chair
¼", T6, G, c. 1940s, D
ot in Back
Karpen Guaranteed Furniture"

179. Dress Maker Dummy
6 1/8", T6, G, 1958, D
Pin Cushion – Slot in Back
National Bk - Little Rock (AR)
Paris dressmaker's model - "Mademoiselle
Pin Money"

180. Coffee Grinder
5 ½", T9, G, c. 1970s, B
Painted Black

181. Rocking Horse
4 ½", T1, S, c. 1950s, E
Winchendon Svg. Bk. (Winchendon,
Ma.) Known as Toy Town

Left: 182. Pentron Record Player
1 5/8", T4, S, c. 1950s, E
"Astra Sente"

Right: 183. Pentron Tape Recorder
1 7/8" T4, S, c. 1950s, E
Top View
Bank is bottom half of tape recorder case

Left: 184. Pfaff Sewing Machine
4 ½", T1, S, c. 1940s, D
Slot in Top
"Dial a Stitch - Pfaff 130"

Right: 185. Cook Stove
4 5/8", T10, G, c. 1970s, B
Painted Black
Six lids on top of stove

Left: 186. Pot Bellied Stove
5 ½", T9, G, c. 1970s, B
Painted Black

Right: 187. Small Suitcase
2 ¾", T39, G, c. 1960s, C
Slot in Top Behind Handle
Southwest Fed Svg & Ln Assn.
"Banthrico Industries - Chicago"

188. Wall Crank Telephone
6 3/8", T8, G, c. 1960s, B
Large Size
St. Louis Telephone Employees Credit Union
Plastic adjustable speaker

189. Wall Crank Telephone
5 3/8", T9, G, c. 1970s, C
Small Size
Des Moines TELCO Credit Union - Des Moines, Ia.
No adjustable speaker

190. Umbrella on Globe
4 ½", T22, G, c. 1970s, C
Plastic – Black
Northwest Security Nat. Bk of Sioux Falls
"Save for a rainy day" on top

191. Umbrella on Globe
4 ½", T22, G, c. 1970s, C
Plastic – Blue

192. Umbrella on Globe
4 ½", T22, G, c. 1960s, B
Plastic – Gold
Winston - Salem Savings & Ln. Assn (NC)

Left: 193. Umbrella on Globe
4 ½", T22, G, c. 1970s, C
Plastic – Red
Iowa State Bk & Tr. Co. - Center Point, Ia

Right: 194. Umbrella on Globe
4 ½", T22, G, c. 1960s, B
Plastic – Silver
First Nat. Bk of Wellston (Mo)

195. Vacuum Cleaner
4 5/8", T1, S, c. 1950s, D
"Filter Queen"

196. Rexair Vacuum Cleaner
5 ¼", T1, S, c. 1950s, D
"Toledo, Ohio"
"Made by Martin Park Co."
"Traps Dust & Lint in Water"

197. Speed Queen Washer
5 3/8", T13, G, c. 1960s, C
First – 1908
Wheel Turns - slot also In back

198. Bulova Watch/Box
3", T6, G, c. 1940s, E
Slot in Back
Bernsteins Dependable Jewelers - Des Moines, Ia.
"National Products Corp."

199. Beehive with Bee
3 ½", T1, G, c. 1940s, D
Slot in Top
Bank of Utah - Ogden, Utah

200. Carved Coconut
2 ½", T3, G, 1962, D
Slot in Top
Small brochure issued with bank

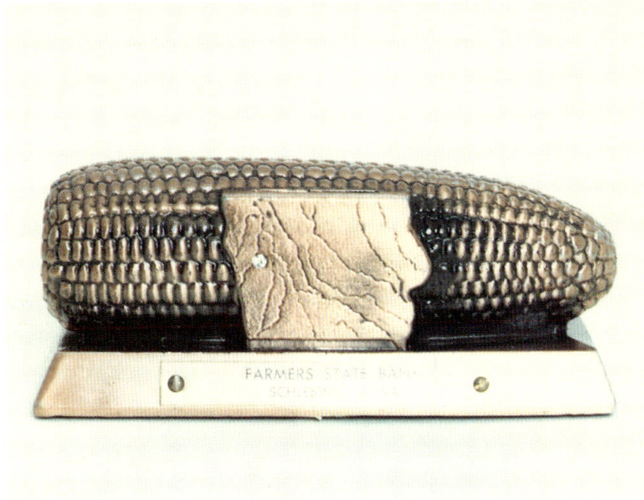

202. Ear of Corn
2 ¼", T8, G, c. 1960s, C
Iowa Map/Stone
Farmers State Bk - Schleswig, Ia

201. Bull on Ear of Corn
4 ¼", T8, S, c. 1970s, C
Animal Chapter Also/131
State Bank of Lizton - Since 1910

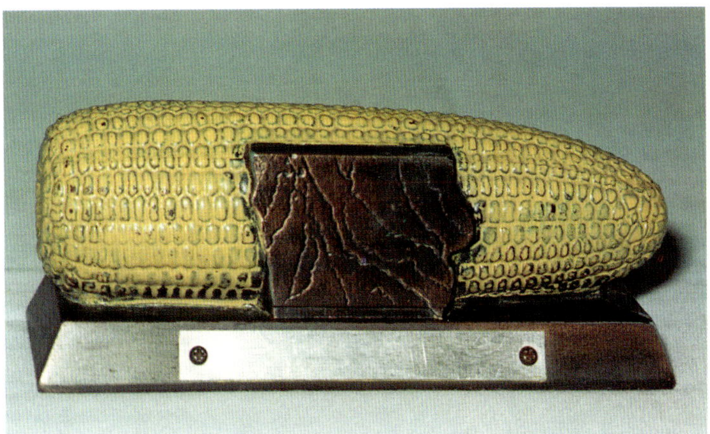

203. Ear of Corn
2 ¼", T4, G, c. 1960s, C
Yellow variation
Metal plate for organization name

204. Idaho Potato – "1st"
3 1/8", T5, S, c. 1960s, B
Slot in Bottom/Top
Idaho First Nat. Bk. - Since 1867

205. Idaho Potato
3 1/8", T3, S, c. 1950s, B
Map of Idaho

206. Idaho Potato
3 7/8", T4, S, c. 1960s, D
Potato on End/Map
Idaho Svg & Ln. Assn.
Pocatello - Boise - Idaho Falls - Twin Falls (Id)

Chapter 7
Trophies

These banks look like trophies that would be given to deserving persons. They are arranged in four categories: disks, eagles, maps, and signs. Within each category, the banks are arranged alphabetically by the sponsoring financial institutions or organizations.

Disk Banks

There are twenty disk banks known, eighteen shown in this group and two others (#68 - Sammy Baugh and #514 Wagon Wheel) in other chapters. Banthrico was prolific in this area. Collectors are familiar with the arrangement by the sponsoring institutions. Fourteen states are represented and Michigan leads the pack with four among the disk banks. A favorite is the Bank of the Commonwealth in Detroit, Michigan, with people and a horn of plenty. "Terra, Economia, and Labor" are written on the bank which shows Detroit buildings, a factory and an airplane in the background. One difficult to find is the Holy Redeemer Church from Flint, Michigan. The back of the bank is just as well made as the front, with a large cross on top of a building and an inscription around the top of the disk ("The Golden Cross"). The Disk with Woman bank (#222) is well executed depicting a woman and electric turbine with factories, trolleys, and homes being powered by electricity. Note that the bank is lettered "Property of the Progressive Building and Loan Association." These banks were loaned to savers to fill and return to the loan company; they were not the property of the savers. It was a common practice in the 1920s through 1950s to indicate that the coin bank belonged to the organization providing it. Banker's Thrift Company and Stronghart both manufactured steel banks with similar messages on them. There are several other disk banks that you will enjoy seeing. Building collectors will especially like disk banks because several buildings are included.

Eagle Banks

Seven eagle banks are known, four are included here and three other eagles are among the disk banks. #225, Eagle on Cube, is a magnificent bird and is a great and beautiful bank! When organizations wanted to convey solidarity, nationalism, and power, the symbolic American eagle was a good choice for their banks. The Eagle on Clock (#227) is a very unusual bank because it could also be used as a planter or a holder for business cards, coins, or paper clips. The receptacle is located on the left side of the bank. The Eagle on the Disk (#224) is also a well-made bank and difficult to find without broken wings.

Map Banks

The four map banks include Arizona, Indiana, Washington, and the U.S.A. One map of Michigan (#219) is in the disk banks category. The Map of the U.S.A. bank is heavy WM painted and probably was made after WWII; it has an "E" rating. Banthrico painted only a few banks and they are difficult to find. Maps appear on banks in other chapters, such as the cob of corn (Iowa) and the Idaho potato.

Sign Banks

Signs come in all shapes (rectangle, cube, oval, round, square, and bell), but they conveyed a message to the saver. The Weather Bell Sign (#235) is ringable; you push the button on the top of the bank. It is a "D" bank. The Puget Sound Bank (#241) has the Seattle skyline and Mt. Ranier in the background. The Ben Franklin Half Dollar, first issued in 1948, adorns the Suffolk Franklin Savings bank (#242). The Utica Square Building Sign bank (#243) will fool you if you don't look at the base; Utica Square is in Tulsa, Oklahoma, not Utica, New York, as might be expected! The Art Deco Deer (#237), with a deer jumping 6-foot fences, logs, and brush, brings back fond memories of hunting.

Two "E" banks (#218 and #232) are in the chapter and eighteen "D" banks are listed. Other banks in this grouping are rated "C." There are no "A" or "B" banks in the trophy section.

207. Disk with Building
5 ¼", T4, S, c. 1950s, C
"Kalamazoo" (MI)
American Nat. Bk & Tr. Co - Kalamazoo (MI)
"Life is good in Kalamazoo"

208. Disk with Building
5 ¼", T4, S, c. 1955, C
"1833 – Iowa's Key City"
American Tr & Svg Bank - Dubuque, Ia
Since 1905

209. Disk with Ship
5 1/8", T4, S, c. 1950s, C
Large Eagle on Reverse
Atlantic Fed Svg & Ln Assn.
Fort Lauderdale, Fl.

210. Disk with People
5 1/8", T3, S, c. 1950s, D
"Terra – Economia – Labor"
Bank of the Commonwealth - Detroit, MI

211. Disk with Building
5 ¼", T3, S, c. 1950s, C
Name on Both Sides
Broadway Nat Bk - Nashville, Tn.

212. Disk with Ship
5 5/8", T1, S, c. 1950, D
Slot in Top
Emigrant Industrial Svg Bk - New York
Founded 1850

213. Disk with Eagle
5 1/8", T1, S, 1964, C
Slot in Top
Fairfield County Tr. Co - Chartered 1889
"75 years of service to the community - 1889-1964"

214. Disk with Building
5 1/8", T4, S, c. 1960s, D
"Meriden, Connecticut"
First Fed S&L Assn - Meriden, CT.
Different building on opposite side - Cheshire, CT.

215. Disk with "1st"
5", T4, G, c. 1974, C
Slot in Top and Bottom
First Nat. Bk. Belleville, Il - 1874

Left: 216. Disk with "1st"
4 ½", T4, S, c. 1963, D
Slot in Top
First National Bank of Chicago - 1863
Bank free for opening svg. acct.

Right: 217. Disk with "1st"
5", T4, G, c. 1960, C
"Since 1870"
First Nat Bk - Seattle, WA.

Left: 218. Disk with Church
5", T4, S, c. 1960s, E
"The Golden Cross"
Holy Redeemer Church - Flint, Mi.

Right: 219. Disk with Map
4 ¾", T3, S, c. 1950s, D
Slot in Top
Michigan Nat. Bk

Left: 220. Disk with Motto
5", T4, S, c. 1953, C
"Established 1903"
Nat. Bk. of Commerce - San Antonio, Tx.

Right: 221. Disk with Building
4 7/8", T4, S, c. 1960s, C
"The City of Roses"
Portland Trust Bk., Portland, Or.

222. Disk with Woman
5 1/8", T4, S, c. 1950s, D
Louisville, Kentucky
Electricity uses, Progressive Bldg. & Ln. Assn.

223. Disk on City Skyline
5 ¼", T4, S, 1954, C
Airplanes in Skyline
Wichita Fed. Svg, Wichita, Ks
1934-1954

224. Disk with Clock/Eagle
6 ¼", T4, S, c. 1950s, D
Slot on Back
Worcester Fed. Svg., Worcester, Ma.
Founded 1877

225. Eagle on Cube
7 ½", T8, S, c. 1965, D
"AM" on Base

226. Eagle on Box
3 7/8", T1, G, c. 1950s, D
Brass Name Plate
First Fed S&L Assn of Gary (In.)

227. Eagle on Clock
6 ½", T1, G, c. 1950s, D
Slot in Top
Home Fed Svg.
Used for planter, business cards & change.

Left: 228. Eagle on Shield
4 5/8", T4, S, c. 1960s, C
South Bend, IN
Nat. Bk & Tr. Co - South Bend (Ir

Right: 229. Map of Arizona
4 ¼", T4, S, c. 1956, C
Slot in Top
1st Nat Bk of Arizona - Est. 188

Left: 230. Map of Indiana
4 ¼", T4, S, c. 1950s, C
"Indiana the Hoosier State"
Hoosier State Bank of Indiana - Hammond, In.

Right: 231. Map of Washington
3 ½", T4, S, c. 1960s, C
Slot in Top
Peoples Nat Bk of Washington - Moses Lake - Ephrata - Soap Lake - Othello - Warden

Left: 232. Map of U. S. A
3 ¼", T13, G, c. 1948, E
Heavy Painted Bank
Slot in top

Right: 233. Colonial People Sign
4 3/8", T5, S, c. 1960s, C
Picture on Both Sides
All American Bk of Chicago

Left: 234. Weather Bell Sign
4 3/8", T6, G, c. 1960s, C
Slot in Top
Bell Fed Svg, Chicago (Il.)

Right: 235. Weather Bell Sign
4 3/8", T6, G, c. 1960s, D
Ringable Bell
"Blue - no change, Red - warmer,
Yellow - colder, snow, rain"
"Grandmother's Club"

: 236. Buildings Sign
8", T9, S, c. 1970s, C
dview Svg & Ln Co.

t: 237. Deer Sign
8", T4, S, c. 1950s, D
)eco Deer
field State Bank - Est.
)

Left: 238. Large Dollar Sign
5", T6, G, c. 1960s, C
Slot in Top
"The Dollar Builder"
Also produced by Unique Coin Service
Inc. - Chicago

Right: 239. Rectangle Sign
3 5/8", T4, G, c. 1960s, D
Slot in Top
First Fed Savings Thrift Club, Michigan
City - LaPorte (MI)
Number "1" on bank

Left: 240. Cube Sign
3 5/8", T4, S, c. 1960s, D
Slot in Top – "PB"
Peoples National Bank

Right: 241. Oval Sign
5 ¼", T4, S, c. 1960s, D
Puget Sound – "PS"
Puget Sound Nat. Bk.

Left: 242. Ben Franklin Sign
3 7/8", T4, S, c. 1950s, C
Half Dollar
Suffolk Franklin Svg. Bk. Boston, Ma.
Inc. 1833

Right: 243. Building Sign
4", T4, S, c. 1950s, D
Slot in Top
Utica Square Nat. Bk. Tulsa, OK.

Left: 244. "Western" Si
3 3/8", T1, S, c. 1948,
Rope Writing
Western Savings Bank
Number 5981 on trap

Right: 245. Oval Sign
2 7/8", T4, S, c. 1955
Slot in Top
Yakima Fed Svg & Ln As
Since 1905

Chapter 8
Geography, Patriotism, and Miscellany

Four banks are in the geography area, two th trees and two with terrains. The Mountain nge Bank has a twenty-mule team on one side d a skier and fisherman on the other side. Is this e Inyo-Mono National Park in California? The lisades Banks is quite attractive with cliffs on one le and a building on the other. The patriotism nks number only five, but they are exciting banks. ne Liberty Bell, quite easily found, was used by any groups. For example, it was a souvenir of the g bell given to the City of Florissant, Missouri, by e Business and Professional Women of Florissant. ne cannon is also a favorite and is easily found. ne Calvary Cap banks that commemorate the nerican Civil War were recently discovered and e rated "D." One is of the South and the other of e North. One has the box and is shown with traps d boxes in Chapter 12. The Continental Drum, ne of the most beautiful and well-molded banks in e Banthrico offering, has flags, the Liberty Bell, d writing on it. It is easy to find, yet desired by ery Banthrico collector.

Miscellany

These thirty four banks are "one of a kind" nks, including several books, barrels, and romeliners. The Victor Adding Machine is a con-rted Banthrico bank, but probably was used as ashtray, paperweight, or container for small desk ms. The bag of money bank is highly desirable th an early trap, came in gold and silver, and st 36 cents. There are seven barrels in this sec-n; the early ones are made of maple wood and e later ones of nickel-plated brass. Also, some nite metal banks followed the brass ones (#260, 261, #264). Banthrico made over a million ooden and brass barrel banks from the 1930s rough the 1950s. They sold well and the spon-ring banks and customers loved them. Banthrico developed a dispensing machine for the sponsors so that customers could acquire barrel banks easily and quickly. The blue metal barrel was part of the "Storkline Series" and it came in blue for a baby boy and pink for a baby girl with an attractive box for the proud parents. The bankers gave these banks to new parents to help them start a savings program for the child. The Whiskey Barrel Bank is white metal with styrene ends and unusual in that the spigot is the non-removable key.

Banthrico made thousands of book banks with hundreds of different advertisements. This group easily could be a specialty category of collecting. Three shown here are metal, plastic leatherette, and one is all plastic (#265). These banks typically advertised financial institutions, appliances, auto dealerships, insurance and real estate companies, social events, and political candidates.

The cash register is a well executed bank dated "1984" (second edition) with "NCRU" on bank, probably standing for "National Cash Register Credit Union." The Chromeliner series of advertising banks (a ton of them were made) comes in many colors; they were started in the 1930s. They, too, advertised many types of companies, organizations, and products. The term "chromeliners" may derive from their construction, they were made of chromed steel (with plastic covers) and were issued as a "line" (series). Banthrico used the term "line" in several of their series banks (Storkline, Staveline, etc.). The oval containers were also made by the thousands and some had no handles and some did. They were made in these shapes so that the saver could easily carry them to the bank in the hand, pocket, or purse and deposit the contents in the bank. The "tin can" bank (Banthrico term) is unusual in that this is the only tin bank known to the author that Banthrico did (actually contracted out). It is covered with pictures of coins and quite desirable. The electric turbine is a tough bank to find and was made for a specific company which limited its production. The flashcube is a bank, photo holder, and paperweight. The Li'l Fort Knox bank has no slot and the bank is opened and closed with a key. The ATM machine and Robot banks seem to go well together. They are 1970s banks that were not plated (with copper or brass), but polished zinc and done in pewter color. This helped Banthrico on the cost. Some of the last white metal banks made by Banthrico in the 1970s and 1980s were done in pewter color to include mascot and building banks.

The safe banks were made of polystyrene (a type of plastic). Numbers 281 and 282 were made for Australian banks. The Safe-T-Safe Bank is a much earlier white metal bank that is on the difficult to find list. The Vault Safe Bank (#284) is clearly an "E" bank. It evidently was the vault for valuables at Fort Vancouver. Torchy and the Pig is probably a gas company piece because of the flame (Torchy), and the pig fits in as a symbol of savings. The treasure chest is a neat bank! The Penny Tree Bank (1953 pennies) is unusual and the molding had to be a difficult one. The pennies were glued on and the bank usually has some pennies missing when found. It was produced in the early 1950s. The base has a recessed area that can be a planter, coin holder or for paper clips in addition to being a bank. Banthrico called it the "Money Tree." The Underwood Typewriter Bank (just secured) is an early bank done by National Products in the late 1930s and continued by Banthrico in the 1940s before and after WWII . They made other company typewriters also.

The hard-to-find "D" banks number eleven in the chapter. The Vault Safe (#284) is a tough "E" bank.

246. Idaho White Pine Log
2 7/8", T4, S, c. 1950s, C
Map of Idaho
Idaho First Nat. Bk - Since 1867

247. Mountain Range
2 7/8", T4, S, c. 1960s, D
20 Mule Team
Inyo-Mono National Bank, Bishop - Mammoth (Ca?)

248. Line of Bold Cliffs
2 ¾", T4, S, c. 1950s, D
Palisades/Hudson River
Palisades Trust Co - Englewood, NJ

Left: 249. Palisades Trust C
2 ¾", T4, S, c. 1950s, D
Back Side – Building/#248

Right: 250. Tree
4 3/8", T4, G, c. 1970s, C
Slot in Top
Minneapolis S&L Assn (MN)
"A good tree to come to for shelter"

Left: 251. Liberty Bell
4", T6, G, c. 1940s, B
Florissant Bank, Florissant (Mo)

Right: 252. Cannon
3 1/8", T9, G, c. 1970s, B
Moveable Wheels

Left: 253. Calvary Cap – Gray
2 ½", T29, G, c. 1950s, D
Southern Army
First Nat. Bk. in St. Louis
Slot on back for wall hanging

Right: 254. Calvary Cap – Blue
2 ½", T29, G, c. 1950s, D
Union Army
Dixon Nat. Bk. "Banthrico Ind."

Left: 255. Continental Drum
3 ¾", T6, G, 1976, B
"Continental Currency – 1776"

Right: 256. Victor Adding Machine
3 3/8", No Trap, S, c. 1940s, D
Key on Panel Opens Top
Slot in back, Converted bank

Left: 257. . Bag of Money
3 3/8", T1, G, c. 1940s, C
Slot in Back of Bag

Right: 258. Small Wooden Barrel
2 7/8", T3, G, c. 1930s, C
Genuine Maple, stained and polished wood
"Acorn Barrel - Half million sold"
"Turned from one piece"

259. Small Metal Barrel
2 3/4", T3, G, c. 1940s, C
Blue Variation

260. Medium Metal Barrel
3 1/2", T3, G, c. 1950s, C
Slot in Top
"Storkline Barrel"
"Save your coins and have a barrel of fun"

261. Large Metal Barrel
4 1/8", T24, G, 1960, B
Banthrico "Koin Keg"
Big Bend Bank - "Webster Groves, Mo"

262. Nickel Plated Barrel
2 3/4", T3, G, c. 1940s, B
Slot in Top
"Staveline Barrel" Brass
Plaza Bank of Commerce - Kansas City, Mo.

263. Nickel Plated Barrel
2 3/4", T17, G, c. 1940s, C
Side Metal Plate
Citizens State Bank - Clare, MI
"Make savings your slogan"

264. Whiskey Barrel/Stand
3 3/4", T27, G, c. 1960s, C
"Ye Olde Coin Cask"
Metal Barrel, styrene end plates

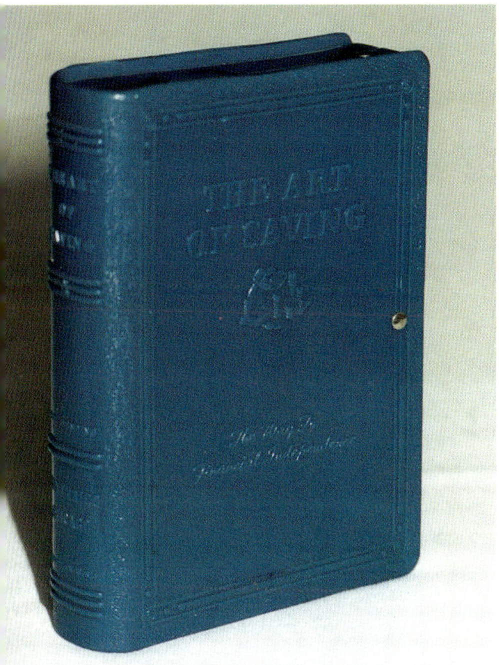

265. Blue Book
4 ½", T30, G, c. 1950s, B
Plastic
Fairbury State Bank, Fairbury, Ne

266. Brown Book
4 ½", T30, G, c. 1940s, C
Slot in Top
Delmar Bank of University City, Mo.

267. Red Book
4 3/8", T30, G, 1940s, C
Slot in Side
"The Art of Saving"

268. Cash Register
5 ¼", T10, S, 1984, C
"Second Edition – 1984"
"NCRU" (National Cash Register Credit Union?)
"Est. April 21, 1937"

269. Chromeliner
4 1/8", T30, G, c. 1940s, B
Green and Red
Phoenix Fed. S&L - Muskogee, Ok.

270. Chromeliner
4 1/8", T30, G, c. 1940s, B
Red and Black
Standard Life Ins. Co of the South - Jackson, Ms.

Left: 271. Chromeliner
4 1/8", T30, G, c. 1940s, B
Red and Black
Jasper County Savings Bank - Newton, Ia

Right: 272. Oval Container
2 ½", T4, G, c. 1950s, B
Nickel Plated
Bank of St. Louis - (Mo)

Left: 273. Oval Container
2 3/8", T3, G, c. 1940s, C
Handle on Top
"1 1/2% Interest on Savings"
Banthrico "Klymax Home Safe."
Also called "satchel" banks

Right: 274. Tin Container
3 7/8", T37, G, c. 1960s, D
Unusual Tin Bank
First Fed. Sav. - Owosso - Chesaning, Mi

275. Electric Turbine Engine
3 7/8", T8, S, c. 1970s, D
"Reliance Electric"

276. Flash Cube
3 ½", T24, G, c. 1960s, B
Photo Holder
Banthrico "Photo Magic"
Liberty S&L, Chicago - Il

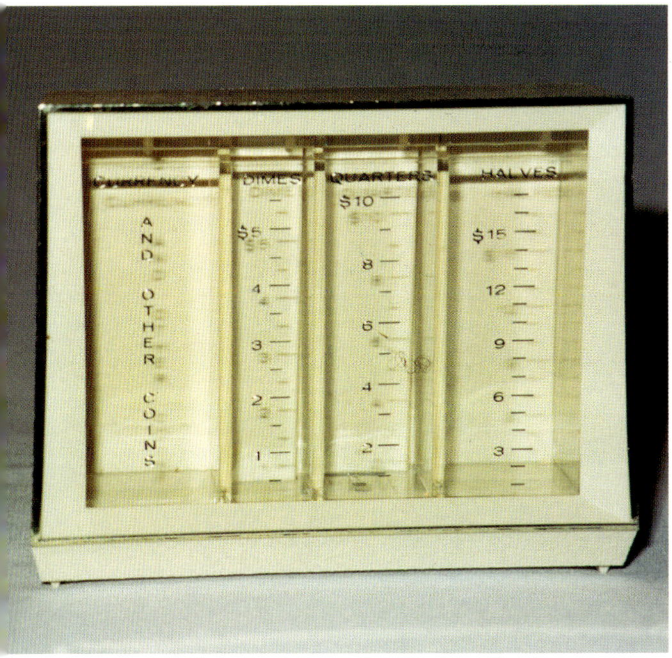

Left: 277. Li'l Fort Knox
4 1/8", T29, G, c. 1960s, C
Has No Slot
Southern Commercial and
Savings Bank - St. Louis, Mo.

Right: 278. "Quickbank" ATM
4 ¾", T10, G, c. 1970s, B
Robot Like Machine

Left: 279. Robot
6 ¼", T6, G, c. 1970s, B
Trap on Back
Tune: "It's a Small World"

Right: 280. Bank Safe
2 ½", No Trap, G, c. 1960s, B
3 Colors – Styrene
"World Book Encyclopedia &
Child Craft"
Slot on back at top

281. Blue Minibank Safe
3 ½", T40, G, c. 1980s, B
Slot in Top
Polystyrene - 4 colors
National Bank Ltd. - Australia

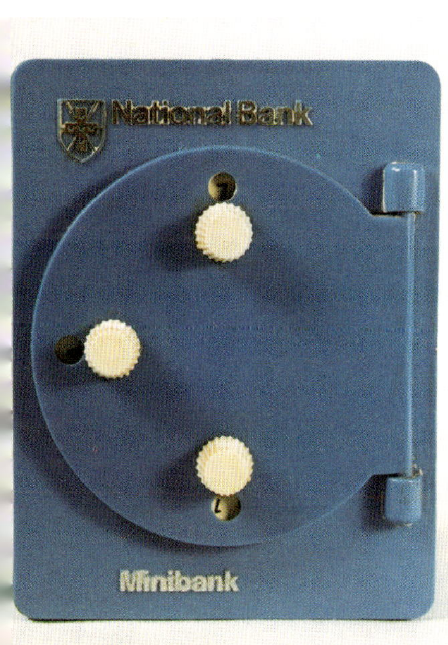

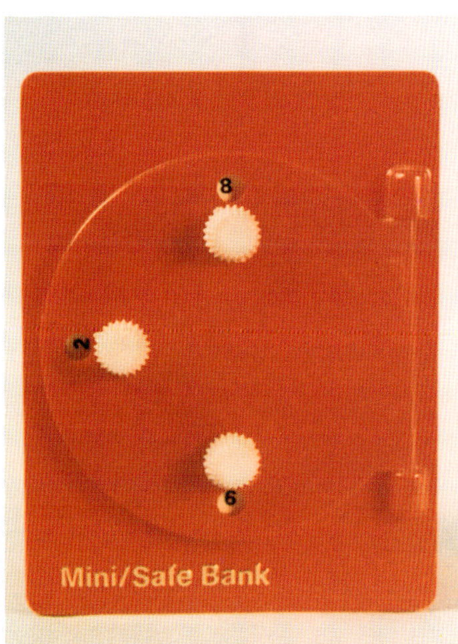

Center: 282. Mini/
Safe Bank
3 ½", T40, G, c.
1980s, B
Red Polystyrene
"Produced in Australia
for Bankers Utilities
Co. Inc."
Slot in top

Right: 283. Safe-T-
Bank
3 5/8", T3, G, c.
1950s, D
Slot in Top
Advertisement plate on
back

284. Vault Safe
4 ¼", T3, S, c. 1950s, E
"Fort Vancouver 1825"
First Fed S&L Assn. - Vancouver, Wa.
Metal name plate

285. Sumshine Bank
2 3/8", T29, G, c. 1960s, B
6 Plastic Coin Areas
National Security Svg & Ln Assn.
Transparent styrene oval bank

286. Torchy and Pig
6 ½", T4, S, c. 1960s, C
Slot in Top
Olympia State Bank - "Save with Torchy"
Torchy - Gas advertising logo?

287. Treasure Chest
5 3/8", T8, G, c. 1970s, D
Full of Old Coins
Treasure Valley State Bk

288. Penny Tree
8 1/8", T1, G, 1953, D
Slot in Top
First Nat. Bk. in St. Louis
(Mo.)
Hammered metal effect

289. Penny Tree
8 1/8", T1, G, 1953, D
Back Side – "1953" Pennies
Second Nat. Bk of
Houston (Tx)

290. Underwood Typewriter
1 3/8", T6, S, c. 1930s, D
Slot in Back
"National Products Corp - Chicago"

Chapter 9
Calculating, Registering and Pocket Banks, and Coin Folders

Calulating Banks

The calculating banks figure out the amount of money that has to be saved daily, monthly, and yearly in order to have a particular amount by a certain date. The front part of the bank, whether a book or a cylinder, has the calculator on it and provides the number tables for the saver. The calculator may have other number tables on it such as insurance rates and amount of insurance for a particular age. The container many also have other advertisements such as Sealy Mattresses and Springs and show the "easy payment plan" for your planned purchase. The calculator wheel can be moved around with the fingernail and set where you want it. This becomes your goal for savings, insurance, etc. If you will note picture #291 at the base of the calculator is "Banthrico, Inc. Chicago." This wording appears on the front and sometimes on the back of the calculating banks. There were several different messages used on the calculators by Banthrico and Bankers Thrift Corporation that were available to the buyer for his clients. The advertisement appeared on the back of the book or cylinder as shown in pictures #294 - #305. These pictures shown are all insurance companies and are from the collection of Tom Vick who specializes in pocket banks. You will note that the insurance salesperson is listed as well as the company and location. Banthrico made several hundred different calculating advertising banks and from those banks came several hundred thousand banks (probably close to a million). Banthrico called the calculating container bank "The Echo Dial Bank." The Insurance Dial Book bank (Banthrico name) (#291) sold for 63 cents each in lots of 500 and the Calculating Cylinder bank sold for 55 cents each in lots of 500.

Registering Banks

Here a coin is needed to activate a calendar date or an amount on the bank. The registering bank was a constant reminder to the saver to practice daily savings. If the saver got a few days behind in his savings plan, he would have to insert several coins to catch up to the current date. The registering banks came on books, chromeliners, and date containers. The chromeliners were also called "streamliners" by Banthrico and seems to be a later term they used without any major change to the shape of the bank but more plastic was used in its manufacture. The coin capacity of the Chromeliner was about $35. Banthrico advertised the Chromeliner as modern styling, beauty, sturdy construction and mechanical perfection. Sometimes you will find a registering bank that does not register properly and that will reduce the rating and value by one or two ratings. The chromeliners were made of steel and a new plastic material and came in six attractive colors: burgundy red, mango green, Monterey blue, Spanish marble, moderne black, and Persian bronze. Banthrico said it was definitely not a mechanical bank. Two of the Chromeliner series were called "Calendar" and "Dimeter." The Calendar Chromeliner bank was an automatic calendar bank and the Dimeter Chromeliner bank was an automatic amount recording bank. The Dimeter was for dimes as the name suggests. The Chromeliner banks had hinged metal key locking traps as indicated by the "T" numbers in the picture captions. Many of the chromeliners had slots in the top of the banks while others had different size slots in the bottom of the banks that would admit up to half dollars and currency. Banthrico did well with the Chromeliner (streamliner) banks starting in the 1930s and continuing into the 1950s making several hundred thousand. They probably changed the name from "Chromeliner" to "Streamliner" since the "in thing" in the late 1940s and 1950s was to "streamline" everything (cars, trains, systems, etc.). The chromeliners were about 72 cents each for lots of 1000. The date container banks evolved from the Chromeliner banks and we see them first in die cast metal in the 1950s (#316 and #317) and then in plastic (#318) in the 1960s. It appears they only registered the date and not the amount, but I may be incorrect on that fact. At least none were observed by the author. The advertisement was on the front and back of the date containers and they had slots for different coin denominations. There is also a picture of the tops of these banks in the Flyer Chapter. The "Date Changer" bank (Banthrico name) would hold a full year savings of dimes ($36.50) or $40 in quarters. Earlier colors were mottled ivory and gold and marble black and gold. Later they came in palomino and surf green metallic colors. The plastic date container has a very unusual trap in the back that slants inward and down. The plastic date container came in ebony and gold, turquoise and gold and coral and gold. The gold was the grille on the bank. They were $1.30 each in lots of 1000 while the die cast earlier date containers were $1.80 each in lots of 1000 (originally $1.35 per 1000, but prices went up). The imprinting cost was usually included in the lot cost if at least 150 banks were ordered.

Arnold Kahne worked as a salesman for the New York Life Insurance Co. in 1936. He sold over 100 life insurance applications in a rather brief time with the use of the Banthrico Calendar Banks (Banthrico name). Banthrico used this information to promote their registering book banks (#306 and #307).

Pocket Banks

The pocket banks are in "a little world of their own." They are quite different from most of the banks that Banthrico made later. Banthrico purchased Bankers Thrift Corporation in 1931 when Bankers was a big pocket bank maker. In addition, Bankers Thrift had purchased the Stronghart Co. in 1929 and they also had been a major pocket bank maker in the 1910s and '20s. Banthrico elected to keep the names Stronghart Co. and Bankers Thrift Corp. on the pocket banks since they were the experts in their field. Therefore, Banthrico got their bank making start with pocket banks and small steel bank buildings (also made by Banker's Thrift Corp.). More will be said about the buildings in Chapter 11. Banthrico finally put their name on a few pocket banks in the late 1930s.

World War II ended the production of pocket banks and they were never made again by Banthrico. Several pocket banks are illustrated in the chapter and they are numbered showing both the front (F) and the back (B). The company who made the bank is listed on the bank's edge and in the picture captions. There is no way of determining the dates of the banks (except #322 and #357); therefore, they could have been made before or after the Banthrico buyout (1931).

Pocket banks are small giveaway banks 2" to 3 1/4" high and usually round, but some were oval. They were easily carried in your pocket or purse and always ready to gobble up any loose change. When filled (usually with dimes due to size) they were taken to the bank and a teller would open the contents and deposit it into your account. The majority of the pocket bank keys were owned by the banks. One bank key (#336) is shown with a pocket bank, illustrating the scarcity of the keys. The pocket banks were advertisement pieces and excellent giveaway banks for the sponsors. They were made of thin steel or tin and celluloid, a material easy to print on so the later pocket banks had printing on them.

Pocket banks all tell a story about the history, cars, trucks, buildings, people and their dress, wording that was used, interest rates that prevailed, and even the colors used when they were made. Some announce a birth with a stork and baby, another shows a couple reflecting upon retirement, one has a ship on the high seas and a date of "1921," many are promoting savings, mortgage rates, interest rates, several show bank buildings and others their capital assets. One bank is lettered, "we loan you this little bank."

Several of the pocket banks are calculating banks. New York and Pennsylvania are the pocket bank leaders with Chicago coming in second. Stronghart, Bankers Thrift and Banthrico probably made over a thousand different advertising pocket banks over a thirty-plus year period. Bankers Thrift Corp. even marketed a "Bankers Dime Bank" dispensing machine for pocket banks. A flyer in Chapter 15 pictures the dispensing machine and the pocket bank (#1065).

One of the pocket banks (#353) is very unusual in that it is 1 3/4" deep. The normal depth of the pocket banks is about 1/2." This bank will need a "big pocket" to carry it.

Coin Folders

The cardboard coin folder banks and cardboard display materials were made by sub-contractors for Banthrico. Eleven banks are shown in this chapter as a sample of another product of the Banthrico Company. There is a flyer in color in Chapter 15 that shows another twenty or so of the coin folder banks. The coin folder contained small pockets where the coins, usually dimes and quarters, could be inserted. The pockets provided a cumulative count of the amount saved and encouraged the saver to continue on their savings goal. See also Chapter 12 for a picture. The coin folders were easily carried in your pocket, purse, or folder and took up little space. Banthrico had stock folders that could be ordered by the thousands and have an organization or company name printed on them. They also made special order folders with the bank building picture (#347, #349, #351) or a name brand article, as shown in picture #344. Favorites were family scenes (#350), western motif (#341), and animals (#348). Banthrico made millions of these banks. They had over two hundred stock folders that were used through the years (1940s - 1980s). The coin folders were ideal World War II banks because they required no metal in the final product.

The Toystalgia Company was not interested in the coin folder banks when they purchased Banthrico in 1985. Therefore, Banthrico's former president Pete Eisendrath retained the coin folder bank business with his present Signs Now business in Chicago, but they do not carry the Banthrico name.

The generic banks are rated as "A" banks and the specific banks are rated as "B." They all seem to be quite plentiful and easily found. It is a good bank specialty field for the beginning collectors with a limited budget.

The custom folders were more expensive than the stock folders because the art department had to develop separate shapes, pictures, wording, etc. requested by the customer. The price for the "Mr. First" (#345, #346) coin folders was $48 per thousand or $39.50 per thousand for 10 thousand. The "Pocket Pig" coin folder (#348) sold for $42 for one thousand and the Covered Wagon (#341) sold for $44 per thousand. The "Santa Dimesaver" coin folder made its debut in 1960 (not shown) and was an instant hit with the Christmas bank clubs and sold for $44 for 1,250. Banthrico made a nativity scene from cardboard during the 1960s, but it was not a bank. Banthrico had all kinds of cardboard display materials for the customer to display their coin folders and other banks and they were often free or available for a very nominal charge.

291. Red Book Calculator
4 ½", T30, G, c. 1930s, C
Slot in Top
Lamar Life Ins. Co

292. Calculating Cylinder
2 5/8", T18, G, c. 1930s, C
Red – Slot in Top
Kansas City Life Ins. Co

293. Calculating Cylinder
2 5/8", T18, G, c. 1930s, C
Green Variation

294. New York Life Ins. Co.
2 5/8", T18, G, c. 1930s, C
"Banthrico Inc. Chicago"

295. Reliance Life Ins. Co.
Pittsburgh (PA)
2 5/8", T18, G, c. 1930s, C

296. Country Life Ins. Co.
Chicago (IL)
2 5/8", T18, G, c. 1930s, C

297. Penn Mutual Life Ins. Co.
Cincinnati, Ohio
2 5/8", T18, G, c. 1930s, C

298. The Mutual Life
Buffalo, New York
2 5/8", T18, G, c. 1930s, C

299. State Mutual Life
New York City, N. Y.
2 5/8", T18, G, c. 1930s, C

300. Massachusetts Mutual
Life Insurance Company
2 5/8", T18, G, c. 1930s, C

301. Equitable Life Assurance
Philadelphia (PA)
2 5/8", T18, G, c. 1930s, C

302. Midland Mutual Life
Akron, Ohio
2 5/8", T18, G, c. 1930s, C

303. Security Mutual Life
Binghamton, N. Y.
2 5/8", T18, G, c. 1930s, C

304. The Mutual Life
Pittsburgh, PA
2 5/8", T18, G, c. 1930s, C

305. Penn Mutual Life
Philadelphia (PA)
2 5/8", T18, G, c. 1930s, C

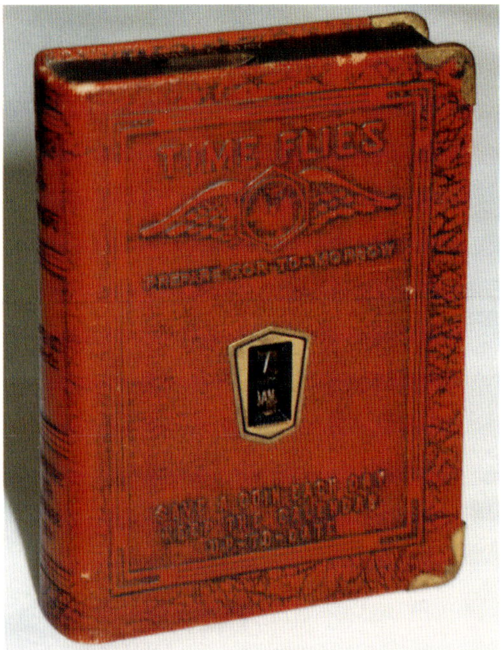

306. Red Calendar Book
4 ½", T30, G, c. 1930s, B
Registers Day and Month
"Time flies - Prepare for tomorrow - Save a coin each day"

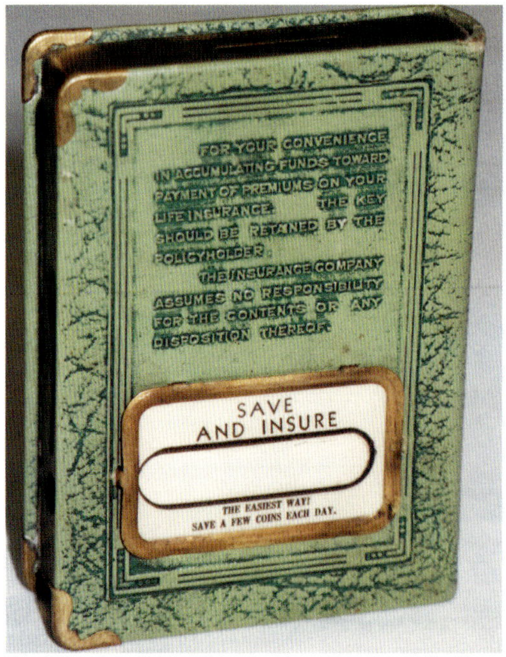

307. Green Calendar Book
4 ½", T30, G, c. 1930s, B
Back Side – Registering Bank
"Save and insure the easiest way"
Slot in top

308. Chromeliner – Blue
4 1/8", T30, G, c. 1930s, C
Registering – Slot in Top
Frankland's (Dept. Store) - Jackson Tn.

309. Chromeliner - Blue
4 1/8", T30, G, c. 1930s, C
Slot in Top
"Frigidaire" with emblem

310. Chromeliner – Blue
4 1/8", T30, G, c. 1930s, C
Slot in Top
First Fed Svgs - Chicago, Il.

311. Chromeliner – Brown
4 1/8", T30, G, c. 1930s, C
Registering
First Mutual Svgs Assn. - Chicago, Il.
"Under state supervision"
"Founded 1882" - Slot in top also

12. Chromeliner - Green
1/8", T30, G, c. 1930s, C
lot in Top
L. Fish Furniture" on fish emblem

313. Chromeliner – Green
4 1/8", T30, G, c. 1930s, C
Registering
Seifers - Hammond, In
"Good furniture for every home"

314. Chromeliner – Red
4 1/8", T30, G, c. 1930s, C
Registering
Capitol Fed Svg & Ln Assoc. - Denver, Co.

15. Chromeliner - Tan
1/8", T30, G, c. 1930s, C
lot in Top
lome Fed S&L Assn of Chicago

316. Date Container
4 1/8", T3, G, c. 1950s, B
Registers Month/Day
Douglas County Bk. of Omaha
- In Benson
"First made in 1955"

317. Date Container
4 1/8", T3, G, c. 1950s, B
Dime/Quarter Slots/Top
Kewanee Fed Svg & Ln Assn.,
Kewanee, Il.

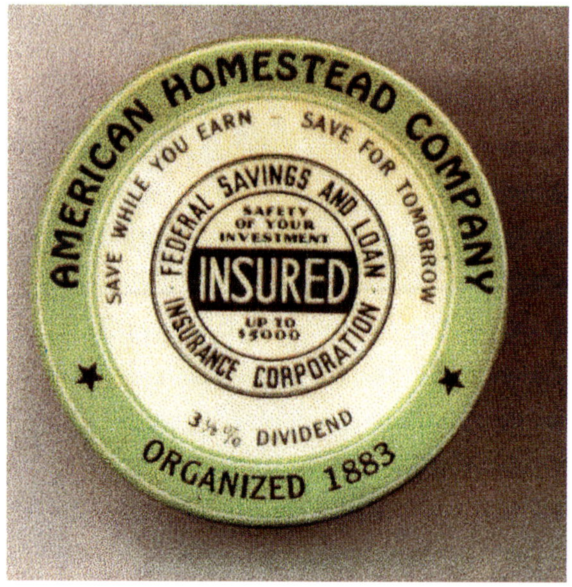

318. Date Container
4 ½", T32, G, c. 1960s, B
Slot in Back
Security Trust Co.

319F. American Homestead Co.
2 ½", T34, S, c 1930s-'40s, C
Banthrico, Inc.

319B. American Homestead
"Banthrico, Inc."
Calculating Bank

320F. American Trust
Le Mars, IA
2 ½", T34, S, c 1920s-'30s, C

321F. Chelsea Exchange
Brooklyn, NY
2 ½", T34, S, c 1920s-'30s, B

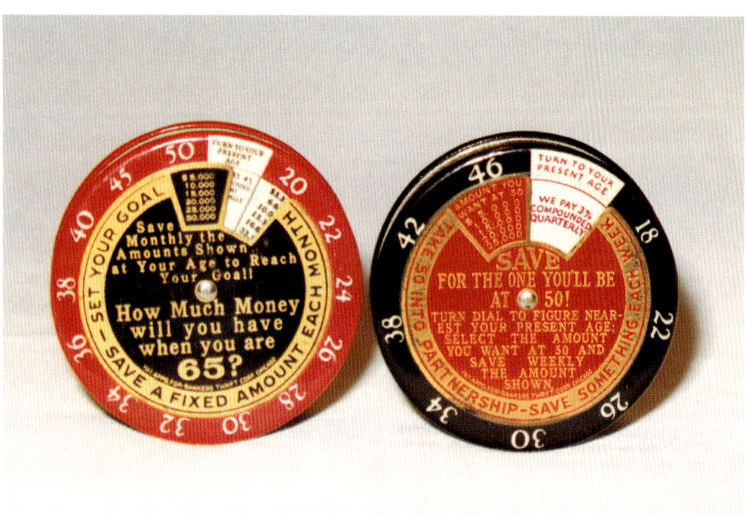

320B. American Trust
Bankers Thrift Corp.
Calculating Bank

321B. Chelsea Exchange
Bankers Thrift Corp.
Calculating Bank

2F. Atlantic Trust
timore (MD)
½", T34, S, 1921, C

3F. Citizens Building
rion, Ohio
¼", T34, S, c. 1910s-20s, D

322B. Atlantic Trust
Stronghart Co.
"1921" Under Ship

323B. Citizens Building
Stronghart Co.

4F. Citizens Banking
ina, Ohio
½", T34, S, c 1920s-'30s, C

5F. Commercial Savings
naha, Nebraska
½", T34, S, c 1920s-'30s, B

324B. Citizens Banking
Bankers Thrift Co.

325B. Commercial Savings
Bankers Thrift Corp.
Calculating Bank

326F. Farmers & Merchants
Stillwater, Minnesota
3 ¼", T34, S, c. 1910s-20s, D

327F. Home Savings
Tipton, Indiana
3 ¼", T34, S, c. 1910s-20s, D

326B. Farmers & Merchants
Stronghart Co.

327B. Home Savings
Stronghart Co.

328B. Jefferson
Gravois
Bankers Thrift
Corp.

328F. Jefferson –
Gravois
St. Louis (MO)
2 ½", T34, S, c 1920s-
'30s, D

329F. Kirkwood Trust
Kirkwood, Missouri
3 ¼", T34, S, c. 1910s-20s, C

330F. First National
Bancroft, Iowa
2", T34, S, c. 1910s-20s, C

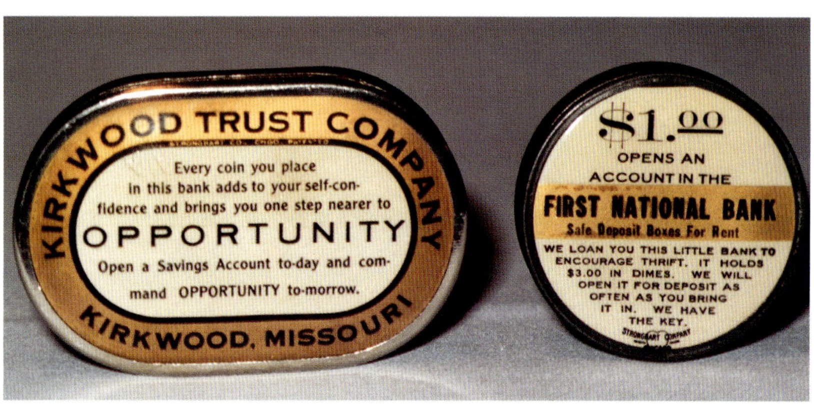

329B. Kirkwood Trust
Stronghart Co.

330B. First National
Stronghart Co.

1F. Merchants & Illinois
ria (IL)
4", T34, S, c. 1910s-20s, D

2F. Park Savings
umbus, Ohio
2", T34, S, c. 1910s-20s, B

331B. Merchants & Illinois
Stronghart Co.

332B. Park Savings
Stronghart Co.

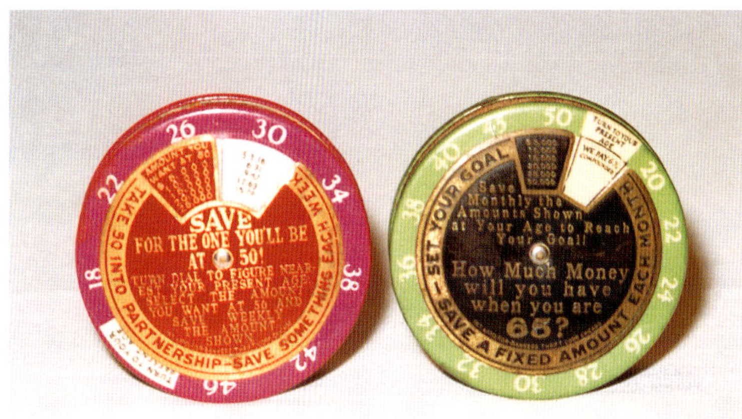

3F. Monroe County
hester, N. Y.
2", T34, S, c 1920s-'30s, C

4F. Prudential Building
at Bend, Kansas
2", T34, S, c 1920s-'30s, B

333B. Monroe County
Bankers Thrift Corp.
Calculating Bank

334B. Prudential Building
Bankers Thrift Corp.
Calculating Bank

335F. National Bank
Hamburg, PA
2 ½", T34, S, c. 1930s-40s, C
Note two feet on bank (not four)

335B. National Bank
Hamburg, PA
"Banthrico, Inc."

336. Older Couple Saving
3 ¼", T34, G, c. 1910s-20s, D
Stronghart Co. /Plain tin back

337F. Pierwszy-Dru
W. Pittsburgh, PA
2 ½", T34, S, c.
1910s-20s, D

338F. Savings Bank
Utica (NY)
3 ¼", T34, S, c.
1910s-20s, D
Note 1910s-20s
cars, trucks and
trolley in picture

337B. Pierwszy-Drugi
Stronghart Co.

338B. Savings Bank
Stronghart Co.

339F. Reserve Building
St. Louis, MO
2 ½", T34, S, c 1920s-'30s, C

340F. Southern Illinois
East St. Louis, Ill.
2", T34, S, c. 1910s-20s, C

339B. Reserve Building
Bankers Thrift Corp.
Calculating Bank

340B. Southern Illinois
Stronghart Co.

Left: 341. Amoskeag Savings
6 3/8", T41, G,19 40-80s, A
Holds $3/Dimes

Right: 342. Bank of America
6", T41, G, 1940-80s, A
Note Mercury Dime

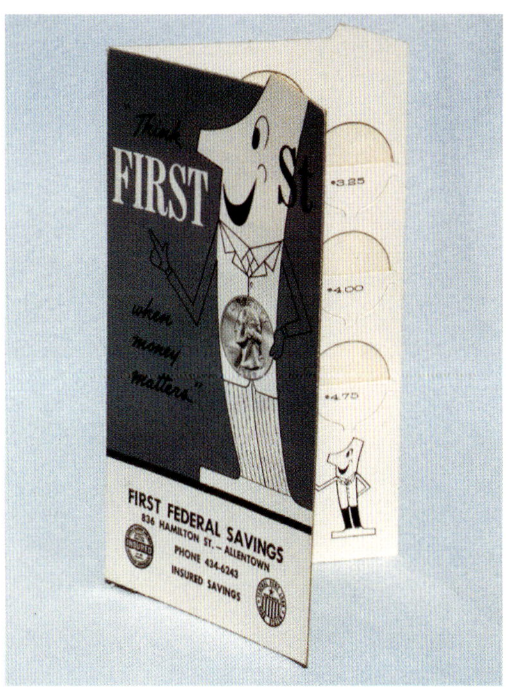

Center left: 343. Bank of America
3 ½", T41, G, 1940-80s, A
Holds $3/Dimes

Center right: 344. Evans Colson Bike
3 ¼", T41, S, 1940-80s, B
Holds $5/Quarters

Bottom left: 345. First Federal – Brown
Allentown (PA)
6 ¼", T41, G, 1940-80s, A
Holds $3 in dimes

Bottom center: 346. First Federal – Blue
Allentown (PA)
6 ¼", T41, G, 1940-80s, A
Holds $5 in quarters

347. Merrill Trust
Eastern Maine
3 ½", T41, S,1940-80s, B

348. Pocket Pig
Jackson, (TN)
3 ¼", T41, G, 1940-80s, A
Note "Mercury Dime" on bank (1916-1945) Dated "1941"
Second Nat. Bk - Jackson (Tn)

349. Saving Bank
Utica, (NY)
6", T41, S, 1940-80s, B
Holds $2.50 in quarters and $1.50 in dimes

350. Western Savings
Buffalo (NY)
6 ¼", T41, S, 1940-80s, B

351. Lincoln National
Fort Wayne (IN)
6 ¼", T41, S, 1940-80s, B

352F. American Trust Co.
St. Louis, MO (?)
3 ¼", T34, S, c. 1910s-20s, D

353F. Northern Bank
Seattle (WA)
3 ¼", T34, S, c. 1910s-20s, D

352B. American Trust Co.
Stronghart Co.

353B. Northern Bank
Stronghart Co.
Unusual 1 ¾" Depth

354F. Drexel State
Chicago (IL)
3 ¼", T34, S, c. 1910s-20s, D

355F. Union Trust
East St. Louis, Ill.
2 ½", T34, S, c. 1910s-20s, C

356F. National
Bank
St. Louis, (MO)
2 ½", T34, S, c
1920s-'30s, C

354B. Drexel State
Stronghart Co.

355B. Union Trust
Stronghart Co.

57F. Ridgley – Farmers
pringfield, Ill.
½", T34, S, 1924, D

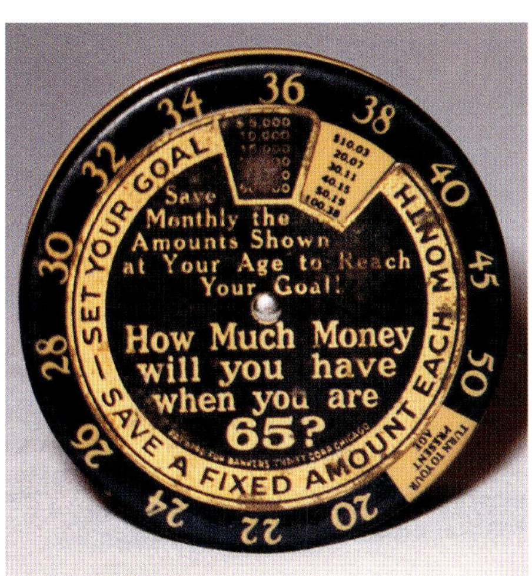

Left: 356B. National
Bank
Bankers Thrift Corp.
Calculating Bank

Right: 357B. Ridgley –
Farmers
Stronghart Co.
"Pat. Oct. 14, 1924"

Chapter 10
Transportation

The transportation banks encompass the land and water, but Banthrico did not include the air transportation by making an airplane bank. They did make an astronaut capsule, and there is a flyer on it in Chapter 15. Many of the land vehicles have moveable wheels and some have non-moveable wheels. The wheel codes are given in Chapter 1, "Reading the Picture Captions." The Banthrico wheel code numbers are presented in chronological order, 1 being the oldest type wheel and number 4 being the newest; 5 and 6 are Valley Casting banks. Also note that the measurement is for the length of the bank, unless indicated otherwise.

When Banthrico purchased National Products Inc. in 1940, they were making tin, steel, and celluloid pocket banks; steel building banks; and lead and white metal banks. When they acquired National Products they inherited a very active and prosperous line of model promotional automobiles and trucks (autobanks). They kept the National Products name on the cars and trucks until about 1949-50 when they added their name with National Products (as a division of Banthrico), and eventually in 1951 they used the Banthrico name only. The 1949-51 were transition years for Banthrico on the promos, so you may find models during these years with either one or both names on the autobanks. World War II (December 1941-August 1945) interrupted the production of white metal autobanks; production was resumed after the war.

Early Autobanks With Rubber Tires

White and black rubber tired white metal promos (autobanks) precede the other transportation banks Banthrico produced. National Products made most of their promos with white rubber tires and Banthrico made their promos with black rubber tires, but there are exceptions during the transition years. Many of the National Products rubber tires had "Firestone" marked on them. Some Banthrico promos today have white replaced tires that are being made by Thomas Toys, Inc. of Fenton, Michigan, tel. (810) 629-8707.

The scale of the promos/autobanks was usually around 1/28th to 1/30th of actual size for National Products and 1/25th for Banthrico, but there were exceptions. The smaller the scale number the larger the autobank. The autobanks scale models were developed from the blueprints of the auto makers such as Cadillac, Chevrolet, Ford, Chrysler, Nash, Studebaker, etc. The promos were issued the year the life-size automobiles were issued and at no other time. They were used to promote dealer car sales and financial institutions' car loans. They were big sellers for National Products and Banthrico, over a million autobanks sold during the 1930s-1950s. Over 100 thousand Chevrolets were sold by Banthrico in 1949. National Products made the first promotional in 1933, a Studebaker President sedan produced for the Chicago World's Fair. It was not a bank. Promos became quite popular and National Products soon had several competitor companies. Banthrico recognized the young thriving company, had the opportunity to purchase it, and expanded the autobank line in the 1940s and '50s.

The colors of the autobanks are the exact colors of the real automobiles. The paint used by Banthrico was the same paint the automobile makers used on their autos. National Products and Banthrico did everything within reason to make the "miniature auto" as realistic as the big auto. The scale, colors, wheels and other details were all considered. The colors available were painted on the roofs of the display models. Some of the Chevrolet colors were oxford maroon, falcon green, Grecian gray, Maryland black, and rodeo beige. The roof was also used for the place of advertisement for the automobile dealers and financial institutions. The imprinting was free if you purchased 100 or more autobanks.

Traps

The traps on the autobanks are mainly a steel metal plate with slot (T6) (oldest), oval key lock trap (T25) (next oldest) and hinged, rectangular key lock trap (T4) (newest). The T6 and T25 traps are mainly National Products and the T4 are mainly Banthrico, but you may see exceptions during the transition period (e.g., T25 on a Banthrico autobank). When Banthrico started using the T4 trap, the name "autobank" on the bottom of the banks gradually disappeared. The name "autobank" on the bottom of the promo was almost always seen with the T25 trap. The T25 oval traps were used only on the promos and then were discontinued. The T4 and T6 are seen on many other banks that Banthrico produced later. In fact, the T4 trap was used the most on the Banthrico white metal banks through the years. On several of the steel base plate autobanks Banthrico added steel metal supports across the plate between the front set of wheels and back set of wheels. This was because of the length of the autobank metal plate and the weight of coins in the bank. These autobanks would hold around $75 in the larger denominational coins.

Banthrico sold autobanks to sponsors throughout the United States and in Latin America and Europe. In fact, a company in Finland purchased the promotional autobank dies of the 1949 Chevrolet from Banthrico and made models for several years.

The promos sold for about $1.50 each if you purchased at least 250. The Futuristic auto wa $1.62 each for 300. The 1952 Buick sold fo $1.47 each for 200 and the 1948 Chevrolet sol for $1.03 each if you purchased 200. The 195 Packard Clipper sold for $1.47 each for 200 o you could buy 12-99 for $1.53 each. The pric went down a few cents for each autobank whe you ordered in larger quantities. The autobanks wer sold by auto dealers and financial institutions fo around $2.50 to $3.00, so they made a profi while advertising their businesses.

Many dealers gave their customers th autobank when they purchased an automobile. The were sent through the mail as well as given out i the showrooms. When a customer was waiting fo their new automobile to arrive from the factory the dealer would provide them an autobank to mair tain interest and promote good will. In the 1950s when the promos were in big demand by collectors Banthrico started selling the promos directly to club and individuals (#1069). They distributed a "Spe cial Collector's List" of "Authentic Scale Models (#1070) which listed the models available, most o which were autobanks. The cost was $2.50 eac with a 20% discount for purchases of 12 or more The flyer is shown in Chapter 15. Banthrico als made the promos available at auto shows and au tomobile museums.

The autobanks are difficult to find today. The are quite fragile (thin white metal) and the pain chips quite easily. You will note that none are rate below a "C," because they are so hard to find. The are so costly that a separate pricing guide was de veloped for the autobanks. Another reason they ar scarce is that model car collectors have many o the autobanks. Also, automobilia collectors focu on anything related to autos and trucks that com from the factories and dealerships, so they collec autobanks, too. The bank collector has lots of con petition for the promotional autobanks.

The autobanks represent fifteen different au tomobile manufacturing companies: Buick, Cadillac Chevrolet, Chrysler, Dodge, Ford, Kaiser, Lincoln Mercury, Nash, Oldsmobile, Packard, Plymouth Pontiac, and Studebaker. At least one model of eac company is shown here. Chevrolet has eleven di ferent models, Chrysler has five, and Ford has five Since the autobanks were painted with actual fac tory paint, there are several of the same model shown in different colors to give the reader an ide of the range National and Banthrico used. Thes promos are miniature automobiles that existed i the United States during the 1930s through th 1950s. When you see these models, you can env sion the cars on the streets of America during thi time frame. The Futuristic car (1955-56) was fror an award-winning auto design probably made by GMC as an experimental car, but never marketed.

Plastic promotionals were introduced in the late 1940s and by the late 1950s Banthrico had reduced their promo line to only an occasional model. They did produce a white metal promo in 1967, but it was not a bank. It was a Willys M-715 U.S. Army stake truck. Plastic had several advantages over the white metal promos: greater detail, color in the plastic, less manufacturing equipment needed, cheaper production costs, easier to ship and more durable to name just some, but the plastic promos usually do not command the prices that the white metal promos do. Banthrico did get into the plastic promos with the Kaiser-Frazer's "Henry J" in 1951. This was the only plastic promo they did. They probably sub-contracted this autobank. Flyers on the Henry J (#1073, #1074) are in Chapter 15 and they are pictured in Chapter 13.

The most difficult to find autobanks are rated "E" and they are the Ford "Woody" Wagon, Kaiser Manhattan, Lincoln Cosmopolitan, Nash Ambassador, Nash Rambler and Studebaker Commander.

Later Metal-wheeled Vehicles

125 banks are pictured in this group: about 60 cars, 17 trucks, 2 fire fighting vehicles, 4 trolleys, a cable car, 10 wagons and stagecoaches, 8 tractors, a couple of race cars, 6 trains and tenders, 8 boats to include a submarine, and some automotive display shelves. These approximate figures include the converted banks, paperweights, ashtrays, and the disk and tractor sign banks.

Many of the moveable wheeled vehicles have the name of the car embossed on the underside of the bank. The year of the car is on the license plate and/or the auto insignia on the grille, so this helps immensely in identifying the Banthricos. But some are not easy to identify, such as the 1912 Packard Town Car, the 1917 Pierce Arrow, the 1900 Pillbox, and the 1908 Cadillac (non-moveable wheels). Many of the identification issues have been solved by consulting Banthrico flyers which picture the banks and provide information about them.

The hard-to-find "D" vehicles and other transportation banks are the Armored car, Pillbox Coupe (#481B), Sailboat on Water, Speed Boat, Train on Base (#504), Trolley with Horse, Trolley with People, Truck Van on Platform, Truck with Logs, Buick Paperweight, GMC Truck, and Sterling Motor Truck. The paperweights were included in the rating since they are often found converted into a bank and definitely a Banthrico transportation product. The really tough "E" banks to locate are the Euclid Earth Mover, Garwood Load Packer, Garwood on Base, John Deere Tractor-1980 (1st Edition), Lincoln Continental-1969, Loraine Crane Truck, Race Car-"1929," Submarine on Base, and Terex Earth Mover. The converted banks are included in the group.

These metal-wheeled vehicles represented over half of the Banthrico sales. Rarely do you find them with boxes. Probably less than half of those made survive today.

Cadillac has four different banks pictured, Chevrolet has six banks, Ford has eleven banks. John Deere has seven banks, Lincoln has three banks and several companies have two different banks. Several banks in this section relate also to the disk and sign banks in Chapter 7.

The Euclid Earth Mover, Loraine Crane Truck, and Terex Earth Mover are all converted banks and difficult to locate. They were first made by National Products (note the black rubber tires) and continued by Banthrico. They are well designed vehicles and were produced as paperweights, desk items and ashtrays.

The Lincoln Continental 1969 and The Submarine are hard to find. The submarine bank was made for one specific bank. The 1929 Race Car is also rare. The Garwood Load Packer appears rarely and The Garwood on Base is similar to the bank and a Banthrico item that is included to help the reader see the different transportation items Banthrico produced. The John Deere Tractor-1981-First Edition bank is hard to find. A large number was issued with boxes and that is how they are usually found. The other five editions were issued in the same number, but are much more plentiful because they are not being kept off the market by collectors.

The 1900 Pillbox Coupe from Germany is an interesting advertisement bank. The Peddler' s Wagon is a rather unique bank that reminds me of the peddler wagons that roamed the Missouri Ozark countryside in the 1930s. The Popcorn Cart (#484) is another interesting and difficult bank to find and is close to a "D" bank. One of the most beautiful and attractive banks that Banthrico made is the Sailing Ship. It is the only bank that Banthrico made in this fashion and has "Old Ironsides" on one side of the bank and the "U.S.S. Constitution" on the other side. That is why it is listed as two different banks. It is rather easy to find and usually reasonably priced.

The Trolley with Horse bank was not for the purpose of transporting horses, but for the horse to pull the trolley up the hill and then the horse got to ride down the hill in the back compartment. If you notice the Trolley with People bank closely, you will see the women dressed with bustles of the early 1900s. The bank was probably made in the 1950s following the war. The two fire-fighting vehicles (#409, #494) are fairly easy to acquire. The Banthrico "Auto Loans" Displays have some different cars on them.

The 1908 Cadillac has four variations shown, one is the same vehicle showing both sides in order to see the man and the woman in the front seat. Two other variations have different inscriptions on the spare tires, and a fourth variation (the newest) has moveable spoked wheels, no people in the car and no spare tire. The Circus Wagon bank is attractive and announces to the parade attendees that "The Circus Is In Town."

The "General" Locomotive and its wood tender belonged to the Western and Atlantic Railroad (W.A.R.R.). It was built in 1855 and became famous in 1862 during the Civil War. James Andrews and his Yankee Raiders stole the locomotive at Big Shanty, Georgia, and headed north toward Chattanooga, Tennessee. The "Great Locomotive Chase" began with William Fuller, conductor of the General, following the stolen locomotive. The 87-mile chase ended eight hours later with the capture of the General. There have been several books written and a movie produced about the chase. The General has been reconditioned by the Louisville and Nashville Railroad and can be seen in that area, moving on its own steam power and providing a link into the past Civil War. The bank is readily available at a minimal cost.

Banthrico made not only the larger and smaller versions of the 1902 Nash (#473, 474) but also one on a base with non-moveable wheels. It was Nash's Golden Anniversary year issue (1902-1952). It was a scale model and paperweight, and not a bank, and is shown in Chapter 15 (#1077).

Those With Music Boxes

In the 1970s and 1980s, Banthrico added musical boxes to their moveable wheeled vehicles. The 1953 Chevy Corvette plays "Try to Remember," the 1930 Cadillac Roadster "Born Free," the 1910 Stanley Steamer "Evergreen," the 1935 Ford Taxi and 1977 Volkswagen "My Way," the 1929 Ford Model "A" plays "The Way We Were," the 1928 Chevy Pickup plays "Feelings," the 1955 Ford Thunderbird plays "Fascination," the Stagecoach plays "Waltz Matilde," the Covered Wagon plays "Home on the Range," and the "General" plays "Around the World in 80 Days." Many of the banks with music boxes were sold directly to gift shops and car museums (#1090, #1091, and #1092).

Original Prices

The prices of some of the vehicles were as follows:
1910 Baker Electric - $2.03 each in lots of 160;
Stage Coach - $1.73 each for lots of 150;
1900 Pillbox - $2.35 each for lots of 160;
1933 Dusenberg, $3.74 each for lots of 160;
1927 Lincoln Brougham, $3.74 each for lots of 160;
1937 Rolls Royce, $3.74 each for lots of 160;
1936 Cord - $3.74 each for lots of 160;
U.S.S. Constitution - $3.59 each for lots of 160;
the "General" - $1.85 each in lots of 150
the tinder $1.25 in lots of 150.

The non-moveable wheel vehicles sold for the following prices:
1908 Cadillac, - $1.73 to $1.85 each for lots of 150;
1926 Model "T" Ford, - $1.73 to $1.85 each for lots of 150;
1924 Chrysler - $1.73 to $1.85 each for lots of 150;
the Speed Boat - $1.79 each for lots of 150.

By 1976, the prices had increased to $3.99 to $4.25 each for smaller lots of 6 to 12. By 1982, the wheeled vehicles had increased to around $5.25 each for lots of 160. We see a steady increase in cost of the white metal banks from the 1950s into 1980s and this is expected with increased costs of metal, manufacturing, wages, tools and equipment and the higher cost of living. Banthrico did a good job of producing a quality product at a reasonable price. Unfortunately, the decreased demand by banks and increased government regulations put Banthrico out of business.

Early American Series

Banthrico developed an Early Americana Series (#1084) in the 1970s which depicts the pageantry of early America. Included in the series were the "U.S.S. Constitution," "Old Ironsides," "Lafayette" Locomotive, Covered Wagon, Stage Coach, Peddler Wagon, Popcorn Wagon, Mack Truck, Ahrens-Fox Fire Pumper, "General" Locomotive steam engine and tender, and steam boat paddle wheeler. Considerable advertisement was done to promote the series and it was quite successful.

358A. Buick Roadmaster Riviera
7 7/8", T25, G, 1, 1952, D
"Buick Eight" on Grille

358B. Buick Roadmaster Riviera
7 7/8", T25, G, 1, 1952, D
"Roadmaster" on Fenders
"Gibraltar Bldg. & Ln Assn., Silver Springs, MD"

358C. Buick Roadmaster Riviera
7 7/8", T25, G, 1, 1952, D
Nice Repaint

359A. Buick Roadmaster Riviera
8", T25, G, 1, 1953, D
"Buick" on Grille

359B. Buick Roadmaster Riviera
8", T25, G, 1az, 1953, D
Nice Repaint

360A. Cadillac – Series 62
7 7/8", T25, G, 1, 1952, C
Four Door Sedan
State Bank & Tr Co., Evanston, Ill.
50th Anniversary Issue - 1902-1952

360B. Cadillac – Series 62
8", T25, G, 1, 1952, C
50th Anniversary Issue
Depositors State Bk, Northville, Mi

360C. Cadillac – Series 62
8", T25, G, 1, 1952, C
"Burlington, Iowa"
Farmers and Merchants Svgs of Burlington, Ia.

360D. Cadillac – Series 62
8", T25, G, 1, 1952, C
"Miami, Florida"
Pan America Bank of Miami - Miami, Fl.

360E. Cadillac – Series 62
7 7/8", T25, G, 1, 1952, C
Illinois Nat. Bk. - Springfield, Ill.
"Finance your next automobile thru our Installment Loan Dept."

361A. Cadillac – Fleetwood
8 1/8", T4, G, 1, 1954, D
60 Special
Note red tail lights

361B. Cadillac – Fleetwood
8 1/8", T4, G, 1, 1954, D
"Akron, Ohio"
North Akron Savg Assn. - Akron, Oh.

362A. Cadillac – Fleetwood
8", T4, G, 1, 1955, D
Drive Up Bank Service - Citizen's Bank

362B. Cadillac – Fleetwood
8", T25, G, 1, 1955, D
Nice Repaint

363. Chevrolet – Styleline Club
7 ¾", T25, G, 1, 1949-50, D
Ruse Chevrolet Co., Columbia, Ill.
"Chevrolet" on grill

364A. Chevrolet – Belair Hardtop
7 ¾", T25, G, 1, 1950, D
Deluxe Styleline
First Nat. Bk of Pueblo (Co)
"Auto and personal loans"

364B. Chevrolet – Belair Hardtop
7 7/8", T25, G, 1, 1950, D
"Louisville, (KY)"
Finance your next car with Bank of Louisville (KY)

365A. Chevrolet – Deluxe Fleetline
7 ¾", T25, G, 1, 1950, D
Note Fender Skirts
"Chevrolet" on front grille

365B. Chevrolet – Deluxe Fleetline
7 5/8", T25, G, 1, 1950, D
2 Door Sedan

366A. Chevrolet – Deluxe Fleetline
7 5/8", T25, G, 1, 1950, C
4 Door Sedan

366B. Chevrolet – Deluxe Fleetline
7 ¾", T25, G, 1, 1950, C
"Windsor Blue" on Roof

367. Chevrolet – Deluxe Styleline
7 5/8", T25, G, 1, 1950, D
Sports Sedan

368A. Chevrolet – Styleline
7 ¾", T25, G, 1, 1950, C
2 Door Town Sedan

368B. Chevrolet - Styleline
7 ¾", T25, G, 1, 1950, C
Club Coupe. Note larger rear window and shorter trun
compared to other Styleline club coupes

369A. Chevrolet – Deluxe Styleline
7 ¾", T25, G, 1, 1950, D
"Oxford Maroon" on Roof

69B. Chevrolet – Deluxe Styleline
¾", T25, G, 1, 1950, D
wo Toned Paint

70A. Chevrolet – 210 Series
5/8", T4, G, 1, 1953, D
irst Fed. S&L Assn. of Rockford (Il)
Where savings pays"

70B. Chevrolet – Deluxe
5/8", T4, G, 1, 1953, D
elmar Bank of University City, Mo.

71. Chevrolet – 210 Series
5/8", T4, G, 1, 1954, D
Moore/Darr Collection
M1509

72. Chevrolet – Belair Hardtop
5/8", T4, G, 1, 1955, C
Coupe

373. Chevrolet – Corvette
6 ¾", T6, G, 1, 1958, D
Nice Repaint
Not sure this is a Banthrico, but wheels, tires and bottom seem to indicate it is.

374. Chrysler – New Yorker
8 1/8", T25, G, 1, 1949, D
Note White Tires

375. Chrysler – New Yorker
8 3/8", T25, G, 1, 1950, D
"Chrysler" on Hood/Trunk
Lincoln State Bk - "Bank by auto"

376. Chrysler – New Yorker
8", T4, G, 1, 1953, D
Newport
"Chrysler" on hood and trunk.
Top repainted

377A. Chrysler – New Yorker
8", T4, G, 1, 1954, C
4 Door Sedan
Easton Taylor Trust Co.

7B. Chrysler – New Yorker
", T4, G, 1, 1954, C
hrysler" on Trunk

7C. Chrysler – New Yorker
", T25, G, 1, 1954, C
ngle Piece Windshield

78A. Chrysler – New Yorker
", T4, G, 1, 1955, D
eluxe Sedan
hrysler" on trunk

78B. Chrysler – New Yorker
", T4, G, 1, 1955, D
ational Bk of Waterloo (Il)
Drive-In teller windows. Large parking lot"

79. Desoto – Custom
", T25, G, 1, 1949, D
ome Fed Svgs - Chicago, Il.
DeSoto" on grille and trunk
Safety since 1885"- Lighthouse seal

380A. Dodge – Coronet
8", T25, G, 1, 1950, C
Showman Motors - Community, Ind.
"Dodge" on grille and trunk

380B. Dodge – Coronet
7 7/8", T25, G, 1, 1950, C
4 Door Sedan

381A. Dodge – Coronet
8", T25, G, 1, 1951-52, C
Dodge on Hood/Trunk
North Shores National Bk. of Chicago
"Drive In Banking"

381B. Dodge – Coronet
8", T25, G, 1, 1951-52, C
Two-Piece Windshield
Peerless Fed S&L Assn., Chicago

382A. Dodge – Coronet
8", T25, G, 1, 1953, C
Citizen's National Bank, Decatur, Il.

2B. Dodge – Coronet
, T25, G, 1, 1953, C
e Repaint

2C. Dodge – Coronet
/8", T25, G, 1, 1953, C
tmeir Bros. Inc., Litchfield, Il.

3. Dodge – Royal
, T4, G, 1, 1954, C
odge" on Trunk

4A. Ford Stationwagon
/8", T25, G, 1, 1949, E
Voody"
ord" on hood front

4B. Ford Stationwagon
/8", T25, G, 1, 1949, E
are Tire on Trunk
paint

385A. Ford – Custom Fordor
8", T4, G, 1, 1950, C
Moore/Darr Collection
Portland Federal Savings
"Where thousands are saving millions"

385B. Ford – Custom Fordor
7 7/8", T25, G, 1, 1950, C
Custom Deluxe 4 Dr. Sedan
Note flat rubber tires from age and heat

386A. Ford – Customline
7 7/8", T25, G, 1, 1953, D
Southwest Bank of St. Louis (Mo)
"For Savings - For auto loans"

386B. Ford – Customline
7 7/8", T25, G, 1, 1953, D
First Nat. Bk of Pueblo (Co)
"Auto and personal loans"

387A. Ford Pickup Truck
7 ½", T25, G, 1, 1953, D
F100 Series

37B. Ford Pickup Truck
½", T25, G, 1, 1953, D
'ord" on Tailgate

8. Ford – Fairlane
¾", T6, G, 1, 1955, C
ote Fancy Hubcaps
icago Fire Dept. Battalion Chief #6
CFD" on front & side. Some were not banks

39A. Futuristic – GMC ?
", T4, G, 1, 1955-56, D
M National Bk. - Abilene, Tx
utobank - Reg. U.S. Pat. Office - Banthrico, Chicago,
A"
turistic and Futuramic- both names used

39B. Futuristic – Experimental
", T4, G, 1, 1955-56, D
'ort Worth, (TX)"
'or my money it's the Fort Worth National Bank"

39C. Futuristic – Not Marketed
7/8", T4, G, 1, 1955-56, D
ate Bank of Wellston, St. Louis, Mo.
'ay bills by check. Open an account today"

390. Kaiser – Manhatten
8", T25, G, 1, 1952, E
4 Door Sedan

391. Kaiser – Manhatten
8", T25, G, 1, 1953, E
"Kaiser" on Front Fenders
"Cardinal Maroon" on roof
Color availability guide

392A. Lincoln Cosmopolitan
8 ½", T25, G, 1, 1951, E
"Cosmopolitan" on Front Fenders
Monmouth Trust & Svgs. Bank, Monmouth, Il.
"Automatic savings on automobile loans"

392B. Lincoln Cosmopolitan
8 3/8", T25, G, 1, 1951, E
Sport Sedan
Third Federal Savings - Ben S. Stefanski, President
"Free parking for patrons"

393A. Mercury – Sports Sedan
7 7/8", T25, G, 1, 1951, D
Moore/Darr Collection

93B. Mercury – 4 Dr. Sedan
7/8", T25, G, 1, 1951, D
Mercury" on Front Fenders
he Exchange Nat. Bk. of Chicago - "The Auto Bank"
30 South LaSalle St.

94. Nash – Ambassador
¾", T25, G, 1, 1949, E
epaint – New White Tires

95A. Nash – Rambler
7/8", T25, G, 1, 1951-52, D
ountry Club Hardtop
"Rambler Country Club" on sides of front fenders

95B. Nash – Rambler
7/8", T25, G, 1, 1951-52, D
Nash Airflyte" on Trunk
ohn Kerbeck and Co. - Philadelphia, PA.
st. 1898 - "Continuous service for over 50 years."
Save your money and drive a Nash Rambler"

95C. Nash – Rambler
¾", T25, G, 1, 1951-52, D
epaint

396. Nash – Rambler
7 7/8", T25, G, 1, 1952, D
Byron Stout - Wichita, Ks.
"Dial 2-1435"

397A. Nash – Rambler
7 5/8", T4, G, 1, 1953, E
Country Club Hardtop

397B. Nash – Rambler
7 ¾", T4, G, 1, 1953, E
"Nash" on Spare Tire

397C. Nash – Rambler
7 ¾", T25, G, 1, 1953, E
"World's Largest Nash Missouri Corp." St. Louis, Mo.
2323 S. Kings Highway - "We sell for less"

398. Oldsmobile – Rocket
8", T4, G, 1, 1953, D
Super 88
National Bk of Commerce - Superior, WI.
"Superior's big strong friendly bank."

99. Oldsmobile – Rocket V8
" 5/8", T4, G, 1, 1954, D
uper 88 Holiday Coupe
indell Trust Co. - 2745 N. Grand Blvd. - St. Louis, Mo

00. Oldsmobile – 88 Holiday Coupe
" 5/8", T4, G, 1, 1955, D
tate Bank of Freeport (Il)

01A. Packard – Clipper
", T4, G, 1, 1953, D
Packard Clipper" on Trunk

01B. Packard – Clipper
", T25, G, 1, 1953, D
awrenceville Nat. Bk (KS)

01C. Packard – Clipper
", T6, G, 1, 1953, D
npainted Zinc/Alum
Packard Clipper" on trunk. "Banthrico" on bottom
late.
as not been electroplated, transition model.

401D. Packard – Clipper
8", T6, G, 1, 1953, D
Copper Plated
Transition model from plated to non-plated models

402. Packard – Clipper
7 5/8", T4, G, 1, 1954, D
"Princeton, Indiana"
Farmer's National Bk. - "FDIC" seal on roof

403A. Plymouth – Cranbrook
7 ¾", T25, G, 1, 1953, D
"Lima, Ohio"
First Fed S&L Assn.

403B. Plymouth – Cranbrook
7 ¾", T25, G, 1, 1953, D
4 Door Sedan

404. Plymouth – Plaza Club
7 ¾", T4, G, 1, 1956, D
2 Door Sedan

05A. Pontiac – Streamliner
", T25, G, 1, 1948, D
eluxe 4 Dr. Sedan

05B. Pontiac – Streamliner
", T25, G, 1, 1948, D
epaint/New White Tires
"National Products" on bottom

06A. Pontiac – Silver Streak
1/8", T25, G, 1, 1949, D
Chicago, (IL)"
irst Fed S&L Assn -
The Place for Savings"

06B. Pontiac – Silver Streak
", T25, G, 1, 1949, D
Westport, Conn."
Vestport Bk & Tr. Co.
A home bank in a town of homes"

06C. Pontiac – Silver Streak
", T25, G, 1, 1949, D
hieftan – "Pontiac"/Trunk
Banthrico," "National Products"

406D. Pontiac – Silver Streak
8", T25, G, 1, 1949, D
"Philadelphia, Penna. "
Marconi Savings & Loan Assn.

406E. Pontiac – Silver Streak
8", T25, G, 1, 1949, D
"Silver Streak" on Front Fenders

407A. Pontiac – Deluxe Catalina
7 7/8", T4, G, 1, 1953, D
Moore/Darr Collection

407B. Pontiac – Deluxe Catalina
7 7/8", T25, G, 1, 1953, D
2 Door Hardtop
Advertisement worn off roof - usually first place to wear.

407C. Pontiac – Deluxe Catalina
7 7/8", T25, G, 1, 1953, D
2 Door Hardtop

408. Studebaker – Starliner
7 7/8", T25, G, 1, 1953, E
Commander – Hardtop Coupe
"Autobank" on trap

409. Ahrens – Fox Pumper – 1924
6", T9, G, 3, 1976, C
Fire Truck

410. Armored Car
6", T6, G, 3, c. 1960s, D
Paper Sticker/Side
Citizens Nat Bk - Springfield, Mo.
Note larger spokes in wheels

11. Baker Electric – 1910
5/8", T8, G, 3, c. 1950s, C
Hastings, Nebraska"
ity Nat. Bank
CNB - With people who care"

412. Buick – 1908
4 7/8", T9, G, 3, c. 1970s, B
"Bloomington, Illinois"
Corn Belt Bank

413. Buick – 1954
8", T10, G, 4, c. 1980s, B
"St. Louis, Mo. " Lemay Bank

414. Cable Car
5 ½", T9, G, 2, c. 1970s, C
"101" on Front

415A. Cadillac – 1908
5 ½", T4, G, 2, c. 1950s, B
Man Driving/Right Side
Note brake on outside of car.
"1908" license plate on some cars, none on others
Limousine

415B. Cadillac – 1908
5 ½", T4, G, 2, c. 1950s, B
Woman/Left Side
North County Bk & Trust Co., Bellefontaine Neighbors, Mo.
"V" Cadillac emblem on radiator

415C. Cadillac - 1908
5 ½", T6, S, 2, c. 1950s, C
"NCNB" on Spare

415D. Cadillac – 1908
5 ½", T6, S, 2, c. 1950s, C
"NB"/Eagle/"Toledo"

15E. Cadillac – 1908
3/8", T9, G, 3, c. 1970s, B
'o People/Spare
ater model, note moveable wheels

416. Cadillac Roadster – 1930
7", T10, G, 4, c. 1970s, B
Luggage Box on Back

417. Cadillac – 1956
8 1/8", T10, G, 4, c. 1980s, B
"V" Cadillac Emblem

418. Cadillac – 1959
8 ½", T10, G, 4, c. 1980s, C
Fins on Rear Fenders
Convertible

419. Chevrolet Sign on Base
6 ¾", T10, S, 2, 1983, B
"1912 – First Chevrolet Produced"
West Side - Auto Employee Federal
Credit Union - Flint, Mi
"First Edition"

420. Chevrolet Pickup – 1928
6", T9, G, 3, c. 1970s, C
Spare Tire in Bed

421. Chevrolet – 1949
7 ¾", T10, G, 4, c. 1980s, B
"Chevrolet" on Grille

422. Chevrolet-Corvette – 1953
6 7/8", T9, G, 4, c. 1970s, C
Year/License Plate

423. Chevrolet – 1955
7 ¾", T10, G, 4, c. 1980s, B
Hardtop Coupe

424. Chevrolet – 1957
6 ¾", T10, G, 4, c. 1980s, B
2 Door Hardtop

25. Chevrolet – 1963
¾", T10, G, 4, c. 1980s, B
rvette Stingray

426A. Chrysler – 1924
6 1/8", T4, G, 2, c. 1950s, C
First Chrysler shown to public.

26B. Chrysler – 1924
1/8", T8, G, 3, c. 1970s, B
are Tire/Fender
st Nat. Bk of McCook, Ne.
blem on radiator

426C. Chrysler - 1924
6 1/8", T4, G, 2, c. 1950s, C
Custom Paint Job

7. Chrysler – 1946
, T9, G, 4, c. 1970s, C
wn and Country

428. Circus Wagon
5", T9, G, 3, c. 1970s, C
Tableau Cage
Two lions on each side of ornate cage.
Edwardsville (Il) Nat Bk & Tr. Co.

429. Cord – 1936
6 7/8", T9, G, 4, 1976, C
Year/License Plate

430. Lg. Covered Wagon
5 ½", T5, S, 2, c. 1950s, C
"Morris Plan/Ca."
"Pioneers of consumer credit"

431. Lg. Covered Wagon
5 5/8", T9, G, 3, c. 1960s, B
Trunk on Side

432. Sm. Covered Wagon
4 5/8", T4, G, 2, c. 1950s, B
Water Barrel/Side
East Alton Svgs & Ln Assn., East Alton, Il.

433. Dodge – 1914
5 3/8", T8, G, 3, c. 1970s, B
"Dodge" on Grille

434. Duesenberg – 1930
6 ¾", T9, G, 4, 1976, C
Spare Tire/Fender
"The Big E" - Erie Savings Bank

35. Euclid Earth Mover
¾", S, 1, c. 1950s, E
ınk Conversion
ıbber tires - paperweight, ashtray, desk item.
robably National Products before Banthrico

436. Ford Omnibus – 1915
5 ¼", T8, G, 3, c. 1970s, B
Depot Hack
City National Bank - CNB - Hastings, Ne.
"With people who care"

37. Ford – 1926
3/8", T4, G, 2, c. 1950s, C
o Year/License Pl.

438. Ford – 1926
5 ½", T9, G, 3, c. 1970s, B
4 Door Sedan
"Missouri Savings" and "1926" on rear spare

39. Ford – 1927
½", T4, G, 2, c. 1950s, C
Haltom City, Texas"
alton City State Bk.
The bank where you come first"

440. Ford – 1929
6", T9, G, 3, 1976, B
Convertible Roadster
Rumble seat

441. Ford – 1934
6 1/8", T9, G, 4, c. 1970s, B
Panel Truck
Spare tire above front fender

442. Ford Taxi – 1935
6 1/8", T9, G, 4, c. 1970s, C
Music Box
Tune: "My Way"

443. Ford Convertible – 1936
6 5/8", T10, G, 4, c. 1980s, B
Spare Tire/Rumble Seat

444. Ford – "Woodie" – 1950
7 ¼", T10, G, 4, c. 1980s, A
Station Wagon
Spare tire on back. "Ford" on grille

445. Ford Pickup – 1953
7 3/8", T10, G, 4, c. 1980s, A
F100 Series

446. Ford – Fairlane – 1955
7 ¾", T10, G, 4, c. 1980s, B
4 Door Sedan

47. Ford – "T" Bird – 1955
1/8", T6, G, 4, c. 1970s, C
underbird
usic Box - Tune: "Fascination"

448. Ford – Mustang – 1965
7", T10, G, 4, c. 1980s, B
"Mustang" on Grille

49. "Garwood Load Packer"
3/8", T5, S, 2, c. 1950s, E
ash Truck

450. Garwood on Base
7 ½", S, 2, c. 1950s, E
Paperweight/Ashtray
Not a bank, a desk item. Similar to bank.
Paper sticker: "Made only by Banthrico Inc. Chicago"

1. "General" Locomotive
, T9, G, 2, c. 1960s, B
eat Train Chase

452. "General" Wood Tender
5", T6, G, 2, c. 1960s, B
"Western & Atlantic R. R. "

453. "The Golden Spike Empire"
5 3/8", T4, S, 4, c. 1960s, C
Spike and Train/Front
Box Elder County Bank - Brigham City, UT.
"Golden spike driven at Promontory, UT. - May 10, 1869"
Map of Utah with spike on back.
Slot in top. Where east met west on Transcontinental Railroad

454. Horseless Carriage – 1902
5 1/8", T5, G, 3, c. 1950s, C
SBCCA – June 1971 – PA.
No "Rambler" on hood, removed - variation
Philadelphia, PA. Convention

455. Jaguar – 1955
7", T9, G, 4, c. 1970s, C
Year on License Plate

456. Jeep – U. S. Mail
5 ½", T10, S, 4, c. 1980s, C
Post Office Eagle Emblem

457. John Deere Tractor – 1980
4 ½", T9, S, 1980, E
"Model A – 1934" – 1st Ed.
John Deere Employees Credit Union

458. John Deer Tractor – 1980
4 ½", T9, S, 1980, C
"Model N – 1916" – 2nd Ed.

459. John Deere Tractor - 1981
4 ½", T9, S, 1981, C
"Model D – 1923" – 3rd Ed.

460. John Deere Tractor – 1981
4 ½", T9, S, 1981, C
"Froelich 1892" – 4th Ed.

461. John Deere Tractor – 1982
4 ½", T9, S, 1982, C
"Model GP – 1928" – 5th Ed.

462. John Deere Tractor - 1982
4 ½", T10, S, 1982, C
Mdl GP Wide Tread-1929 – 6th Ed.

463. John Deere Tractor – 1986
4 ½", T8, S, 1986, C
"1911-1986 – 75th Anniversary"
Mowing Tractor "John Deere Horicon Works"

464. Lafayette Steam Engine
5 5/8", T9, G, 2, c. 1970s, C
"B&O 1837"
Baltimore and Ohio

465. LaSalle – 1934
7", T9, G, 4, c. 1970s, C
2 Door Coupe

66. Lincoln Brougham – 1927
", T8, G, 4, 1976, B
'hauffeur Section

467. Lincoln Continental – 1941
7 ¼", T9, G, 4, c. 1970s, C
Spare Tire/Trunk

68. Lincoln Continental – 1969
", T9, G, 4, c. 1970s, E
lark III "CM" on rear plate
las 5 screws to hold long base plate.

469. Lorain Crane Truck
7 3/8", G, 1, c. 1950s, E
Converted Bank
Banthrico advertising scale model truck
slot on top - trap in bottom

70. Mack Truck – 1906
", T8, G, 3, 1975, A
Hastings, Nebraska"
ity Nat. Bk.

471. Marmon Wasp Racer – 1911
7", T9, G, 3, 1976, C
"Florissant, Mo. "
Florissant Bank

472. MG – 1953
6 5/8", T10, G, 4, c. 1970s, C
Note Door Opening

473. Nash Rambler – 1902
6 ¼", T6, G, 2, c. 1950s, C
Large Version
Belleville Nat Svgs. Bank - Belleville, Il.
Variation - slot in bottom of white metal next to plate.

474. Nash Rambler – 1902
5 1/8", T8, G, 3, c. 1970s, B
Small Version
St. Johns Bk & Tr Co., St. Johns, Mo.
Steering rod

475. Oldsmobile – 1906
5 ¼", T9, G, 3, c. 1970s, B
Lanterns/Lights
Raised back seat. "Oldsmobile" on grille

476. Oldsmobile – 1954
7 5/8", T10, G, 4, c. 1980s, A
Autobank Mold

477. Packard Town Car – 1912
5 ¼", T9, G, 3, c. 1970s, C
Tool Box/Running Board

78. Packard V12 – 1937
", T9, G, 4, c. 1970s, B
ote Door Opening

479. Peddler Wagon
5 1/8", T9, G, 3, c. 1970s, C
"Pots and Pans" "Dry Goods"

80. Pierce Arrow – 1917
", T4, G, 3, c. 1970s, C
ide Curtains

481A. Pillbox Coupe – 1900
4 ¾", T4, G, 3, c. 1950s, C
Soft Top
Collinsville Bldg. & Ln Assn., Collinsville, Il.
"Since 1885"

81B. Pillbox Coupe – 1900
¾", T4, G, 3, c. 1950s, D
ermany/Advertising
Big Red Baron" - German Cross on hood.
Genuine German Parts"

482. Pontiac – 1926
5 7/8", T9, G, 3, c. 1970s, C
Spare Tire/Trunk
Pontiac Indian emblem on radiator

483. Pontiac GTO – 1964
7 7/8", T10, G, 4, c. 1980s, A
Hood Scoops

484. Popcorn Cart
4 ¼", T10, G, 3, c. 1970s, C
Handles on Sides
First Nat. Bk.

485. Popcorn Wagon
4 ¾", T9, G, 3, c. 1970s, C
Note Man Inside
"Popcorn - Peanuts - Cold Drinks"

486. Race Car – "1929"
5 7/8", T4, S, 2, c. 1950s, E
Note Building
Speedway State Bank - Speedway, In.

487. Rolls Royce – 1937
7", T9, G, 4, 1976, B
Phantom III Coupe
Spare tire in fender

488. Sailboat on Water
5 1/8"H, T9, G, c. 1970s, D
"Est. 1852"
Westport Bk & Tr Co.

489B. Sailing Ship
5 ¾"H, T9, G, 1976, B
"U. S. S. Constitution"
Backside / 489A

489A. Sailing Ship
5 ¾"H, T9, G, 1976, B
"Old Ironsides"
Introduced in 1976

490. Speed Boat
5 ¾", T4, G, c. 1960s, D
Suffolk County Fed Svgs & Loan Assn.

491. Stagecoach on Base
4 5/8", T4, G, 2, 1960, C
Luggage on Top
Capitol Fed Svgs - Denver (Co)

492. Stagecoach
5", T10, G, 3, c. 1970s, A
Driver's Seat

493. Stanley Steamer – 1910
5 ½", T9, G, 3, c. 1950s, C
"Stanley" on Front

495. Steam Engine Train
7 1/8", T4, S, 2, c. 1950s, C
With Log Tender
Railroadmen's Federal - 1887, Indianapolis (In)
"Safety for savings"

494. Steam Engine
5 3/8", T9, G, 3, c. 1970s, B
Fire Engine Pumper
Vertical boiler

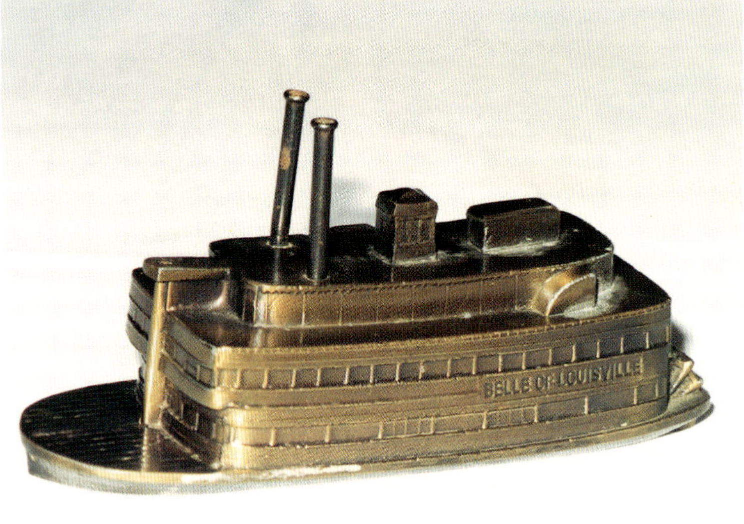

496. Steamboat/Paddle Wheeler
6 ¾", T6, G, c. 1970s, C
"Belle of Louisville"

497. Steamboat/Paddle Wheeler
6 5/8", T9, G, c. 1970s, B
"Florissant, Mo. "

498. Steamboat/Paddle Wheeler
7 5/8", T9, G, c. 1970s, C
"Robt. E. Lee"
Music Box - Tune: "Moon River"

499. Studebaker – 1904
4 5/8", T9, G, 3, c. 1970s, C
Steering Rod

500. Studebaker – 1958
7 5/8", T10, G, 4, c. 1980s, C
"Golden Hawk"/Fenders

501. Submarine on Base
8", T4, S, c. 1950s, E
"Confederate Hunley"
"Sank Federal gunboat
Housatonic, Feb. 17, 1864"
Citizens and Southern National
Bk of South Carolina

02. Terex Earth Mover
", G, 1, c. 1950s, E
onverted Bank
ational Products/Banthrico
dvertising Product. Rubber tires

503. Tractor
5 ½", T10, G, 2, c. 1970s, A

504. Train on Base
7 1/8", T9, S, 2, c. 1970s, D
"1902 – 1929"
Ohio Electric Railway - Interurban
Columbus - Newark – Zanesville

505. Trolley on Incline
5 ¼", T4, G, 2, c. 1960s, B
"San Francisco (CA)"
Citizens Federal Svgs. Oakland -
Palo Alto - "75th Anniversary
1885 - 1960"

506. Trolley with Horse
5 ½", T4, G, 2, c. 1960s, D
"Englewood, Colorado"
First National Bank
"Cheerelyn No 2" - both sides

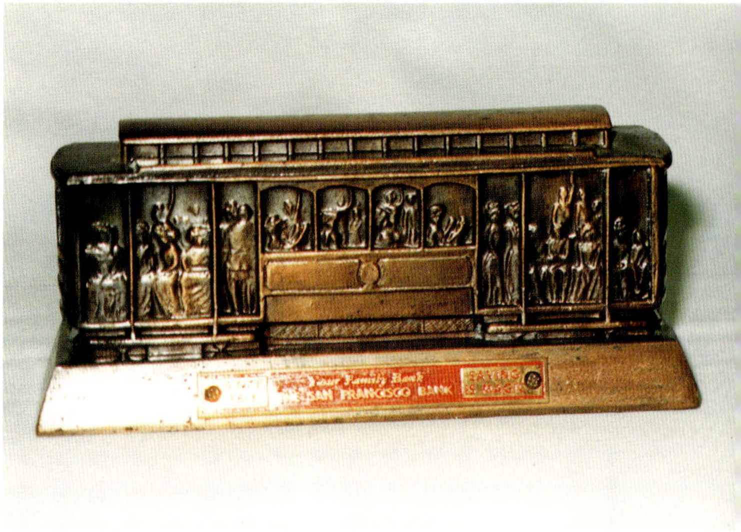

507. Trolley with People
6 5/8", T1, G, 2, c. 1950s, D
Women in Bustles
The San Francisco Bank -
"Since 1868 - Your family bank"

508. Truck Van on Platform
4 ¼", T4, G, 2, c. 1960s, D
"Banker's Dispatch Corp. "

509. Truck with Barrels
6 1/8", T9, G, 3, c. 1970s, B
Music Box: "Edelweiss"
Banthrico "Beer truck"

510. Truck with Logs
6 5/8", T4, G, 2, c. 1950s, D
"Cowlitz County"
Bank of Cowlitz

511. U. S. Mail Truck – 1919
5 ½", T8, G, 3, c. 1960s, B
"Parcel Post"
Lemay Bank - St. Louis, Mo.

Top left: 512. U. S. Mail Truck – 1919
5 1/2", T8, G, 3, c. 1960s, B
Green Painted Variation
"U.S. Mail" paper sticker
on side

Top right: 513. Volkswagen Beetle – 1977
6 1/2", T9, G, 4, c. 1970s, C
Year on License Plate

Center left: 514. Wagon Wheel on Base
5 1/8", T4, G, 2, c. 1950s, C
Covered Wagon/Oxen
Peoples National Bk of Washington
slot in top

Center: 515. Bank "Auto Loans" Display
19"H x 17"W, c. 1970s
Info on Both Sides
1915 Chevrolet
1910 Stanley Steamer
1926 Model "T" Ford
1917 Pierce Arrow

517. Chevrolet Roadster – 1915
5 3/8", T4, G, 3, c. 1960s, C
Soft Top

516. Bank "Auto Loans" Display
19"H x 17"W, c. 1970s
Info on Both Sides
1915 Chevrolet
1902 Horseless Carriage
1926 Model "T" Ford
1910 Baker Electric

518. Buick Paperweight – 1952
7", G, 2, c. 1950s, D
Converted Bank
Probably first made by National Products then Banthrico

519. GMC Truck
6 ¾", S, 2, c. 1940s, D
Paperweight/Ashtray
Paper "Banthrico" sticker on bottom. Not a bank.

Center left: 520. Sterling Motor Truck
6 ½", S, 2, c. 1930s, D
Chain Driven/Paperweight
Also an ashtray
Not a bank

Center right: 521. Baker Electric – 1910
4 5/8", T5, G, 3, c. 1950s, C
Custom Paint Job

Bottom left: 522F. Banthrico Display – Front
Cardboard/Provided with Purchase
Measurements: 9" x 13 3/4"

Bottom right: 522B. Banthrico Display – Back
Transportation Bank
Men of History Banks

Chapter 11
Architecture

The 236 architectural banks comprise five sec-
ɔns: Buildings, Monuments, Statues, Towers, and
nnels. The buildings section is the largest with
)8 banks. The banks are mainly white metal, but
ere are several early lead ones and they are iden-
ied in the captions. There is one plastic mechani-
l drive-in bank (#572), it is the only mechanical
nk in the chapter. One composition bank (#533)
d several steel generic banks appear that Bank-
s Thrift Corporation and Banthrico produced. The
eel banks were continued through the 1930s and
rly 1940s (prior to WWII). There are eight plas-
: generic building banks made during WWII and
llowing the war through the 1940s and early
)50s. Differently colored ones and different styles
e pictured to indicate the variety of plastic build-
g banks Banthrico created. Banthrico contracted
ıt the plastic banks to US companies and the com-
ɔsition banks to Japanese companies.

The architectural banks are listed alphabetically
/ the sponsoring financial institutions and com-
nies. A slight departure from that order occurs in
e brief monument and statue sections where they
e listed alphabetically by their subjects. The infor-
ation in the captions is spelled as it appears on
e banks, e.g. Il., Ill., or Illinois.

The majority of the banks are specific, which
akes them difficult to find and expensive. Specific
nks are made for just one organization and they
e issued in smaller numbers than generic banks.
ere are a few generic building banks shown be-
use they are among the Banthrico products.

There are nine lead banks in the chapter, all
rly (1930s and 1940s). They appear to be lead,
ve excellent details, and are heavy. They include:
ontinental Assurance Co. (#565),
rst Federal Savings and Loan Assn. of Detroit (#592A),
rst National Bank of Danville (#607),
nchburg National Bank and Trust Co. (#651),
arion County State Bank (#653),
atick Federal Savings Building (#663),
ational City Bank of Evansville (#667),
cond National Bank, Nashua, N.H. (#694),
d Serial Federal Savings and Loan Assn. (#697).

There are probably others that will be discov-
ed. Banthrico also made white metal banks from
ese same molds and a good example is pictured
ith the First Federal Savings and Loan Assn. of
etroit (#592B). The white metal casting varia-
ɔns of these lead banks would be rated as "D"
nks, while the lead banks are usually rated "E"
"F." It appears Banthrico made the lead banks
ior to WWII and none thereafter. Zinc is a much
eaper metal than lead and easier to work with in
e molds.

Forty-four states and two countries are represented in the chapter. Illinois leads with 57 banks pictured. Twenty-nine of the Illinois banks are from Chicago, so it is by far the leading city. Texas was second with 17 banks, and close behind was Indiana with 16. Florida had 12 banks, Massachusetts had 11, and New York had 9. Others with five or more are Colorado (8), California (7), Iowa and Pennsylvania (6) each, and Ohio (5). All the other states have less than five banks shown. There is one bank of a building in Mexico (#536) and one bank of a building in Puerto Rico (#597). Only one white metal building bank was done of a Missouri building (#696). Minnesota also had only one white metal building bank (#611) and two different WM towers (#736, #740). Wisconsin has two WM banks (#639, #719).

Rating the Banks

Rating the architectural banks was done by the "Banthrico Board" composed of twelve collectors from around the country and the author. Dort Brown's two books, *Souvenir Buildings, A Collection of Identified Miniatures*, Volumes I and II; Lance Oldham's pamphlet on *Banthrico/Toystalgia/Valley Casting Co. - Building/Monument/Landmark Bank Replicas*; and Margaret Majua and David Weingarten's *Monumental Miniatures* (MM) book all were used as reference material in developing the bank ratings. *Monumental Miniatures* uses a different rating system (I-V) and they rate all types of building items (ashtrays, bookends, cigar cutters, clocks, inkwells, lighters, lamps, penstands, matchbox holders, and pencil holders to name some) that include coin banks, so that was taken into consideration when using the information. Dort tells how many Banthrico banks had been produced when her books were written (1977 and 1979). The ratings are subjective, yet they should be fairly accurate.

Dating the Banks

Dating the architectural banks is one of the most difficult assignments in the chapter. Some of the bank flyers in Chapter 15 have dates, but most do not. The locations of the Banthrico Co. on the flyers was an indicator of the year(s) the bank was made. Banthrico was in three locations in Chicago: Lake Street (1931 - 1940s), Des Plaines Street (1950s - 1960s) and Roosevelt Road (1970s through 1985). The exact date of leaving Lake Street and moving to Des Plaines Street is not known. Brown's book was helpful in determining when some of the Banthrico banks were produced. *Monumental Miniatures* was helpful in establishing circa dates for the banks. Some of the banks have the date on them, but most do not have a date. The traps can be helpful in establishing circa dates, but a bank may have had three or four different traps over the years if it was reissued. Chapter 12 covers trap variations in greater detail.

Use of Parentheses

When information is provided in parenthesis in the captions, that information does not appear on the bank but was gleaned from additional research. The locations of buildings banks may vary here as compared with *Monumental Miniatures* and other books because new information was available.

Toystalgia and Valley Casting (VC) banks are covered in Chapter 14. There should be a very definite division between Banthrico banks and Valley Casting banks because the casting is different, the plating is different, the ratings are different, the companies are different, and the age of the banks is different.

There are 90 "D" banks in this group (around 40%), 25 "E" banks (around 11%), and 4 "F" banks. The "F" banks are Banco Atlantida (#536), First Federal Savings and Loan Assn. - Puerto Rico (#597), Marion County State Bank (#653), and Rapides Bank and Trust Co (#685). The ratings may need some "fine tuning" as the years roll by. These architectural banks are difficult to locate now.

The SBCS/SBCCA Banthrico Building Banks List (developed by Margaret Majua, Dixie Trainer, and Barry Hoffman and first printed in the Souvenir Building Collector in September 1996) has several entries that can now be clarified or corrected. The numbers given are the SBCS List.

The Citizens National bank of Freeport, N.Y. (#33) is a Rehberger bank and is probably not a Banthrico also. It is shown in MM (#1116).

The First National Bank, Davenport, Ia. (#77) is an Art Metal Works bank and probably not a Banthrico also. It is #874 in MM and #832 in Don Duer's *Penny Banks Around the World*.

The National Bank and Trust, Indianapolis, In. (#124) is the same bank as Merchant's National Bank and Trust Co., Indianapolis, In (#116). One name is on one side of the bank and the other name is on the other side of the bank.

All other banks on the SBCS/SBCCA list are included in this chapter or Chapter 14.

The following are attributed Banthrico building banks that have not been located, nor are they pictured here:

First National Bank (or Federal S&L) of Gary, In.
Griswold Bank of Chicago
Harrisburg Federal Savings (Harrisburg, Pa.)
Next Frontier Tobacco Co.

Other Banthrico building banks shown on the Banthrico advertising flyers probably exist (unless Banthrico did not issue them, and that is not likely since they appeared on flyers) and they are not pictured in the book but are shown in Chapter 15 (#1097 and #1104). They include:

National Bank of Wilson

Avenue State Bank (104 No. Oak Park Ave)

"The Bank on the Corner" (Danley Ave.)

The Willman (or Millman) State Savings Bank, Chicago, (South Park Ave. and 1110 Stacey)

Straus Building, Chicago (which appears to be almost identical to the Continental Assurance Co. building (#565) except the entrance and the first floor windows, also see MM #1177 - Straus Building - Rehberger).

A building has "589 Broadway" on front and "National Bank of _" (can't read) on top. These are definitely 1930s and 1940s banks, very difficult to find, and most are probably lead.

Vulcan Tower is pictured on a flyer. It was probably done in the 1960s to 1970s (#1017).

Banthrico and Rehberger made several of the same building banks. The Straus Building was just mentioned. The Jefferson Standard Life Ins. Co. (#642) was made by Banthrico and also by Rehberger (#823 in Duer's book). The Rehberger bank is identified on the front while the Banthrico bank is not identified. It is difficult to figure out which bank came first, but my guess is the Banthrico and then Jefferson Standard decided to add more wording and Rehberger got the contract. Another building made by both Banthrico and Rehberger is the Dollar Savings Bank (#570). It can be seen as a Rehberger in MM (#1125). The Leon Federal Savings and Loan Assn. (#647) was also made by Banthrico and Rehberger (MM #1146). The Syracuse Savings Bank (#707) was made by Rehberger (MM #1176) in the 1930s and by Banthrico in the 1960s. It appears that Banthrico and Rehberger did some sharing of molds; they were both located in Chicago.

The Coast Federal Savings and Loan Assn. - Los Angeles, CA, is a very interesting building bank. It is the rabbit sitting by a bank building, probably a Banthrico bank. The T3 trap looks like the other T3 Banthrico traps. My Coast bank has a "Banthrico" stamped steel key attached that fits the trap perfectly. Duer lists it as American (#833) and MM has it identified as USA (#30). Banthrico did numerous mascot and animal banks, so they certainly had the capability of doing a rabbit bank like this.

Banthrico used the term "Souvenir" in the 1950s. There is a flyer in Chapter 15 (#1112) that uses the language "Ornamental Souvenir and Savings Bank" which pictures the First National Bank and Trust Co. of Tulsa (#617). The term was a harbinger of things to come.

The Bangor Savings Bank (#537) was first produced by Kronheimer and Oldenbusch (MM #1083) in 1912 in lead, later by Banthrico in the 1980s (or maybe earlier) in white metal, and has now been produced by Valley Casting in 1992 in white metal. Maybe Banthrico bought the K&O Bangor Savings mold in 1942 when K&O closed and recast the bank (and maybe others) after the war. Banthrico purchased several companies during this time. (See Penny Bank Post article on Kronheimer & Oldenbusch, December 1998, p. 14-16, by Majua/Weingarten). The Banthrico casting is identical to the 1912 K&O casting and has the same dates "1852-1912" on the bank. The Valley Casting bank casting is identical to the Banthrico casting and has the different dates "1852 to 1992" stamped on the bottom. Valley Casting purchased all of the Banthrico molds in 1985 when Banthrico closed which probably included the Bangor Savings Bank. The lead casting by K&O has much greater detail than Banthrico, and the Banthrico detail surpasses the Valley Casting detail. Lead building banks have greater detail than white metal building banks because lead is a finer metal than white metal (zinc and aluminum). Lead is also several times more expensive.

Variation Banks

Several bank variations are pictured here. Three variations of the Improvement Federal Savings and Loan Assn. in Aurora, Il. are illustrated. The first (#637A) has no "Federal" in the title, which is engraved on the top in the metal. The second (#637B) has "Now Improvement Federal" painted on the back of the bank. The third (#637C) has "Improvement Federal Savings and Loan Association" plus "Aurora Illinois" and the bank title is engraved on the roof.

The American Fletcher National Bank of Indianapolis, In. (#528A) also has three variations. The first is Fidelity Bank and Trust Co. (#585). The second is American Fletcher National B. & T. (#528A) with a metal plate (covering the Fidelity Bank and Trust engraving). The third is the engraved American Fletcher National Bank (#528B).

The Exchange National Bank of Colorado Springs, Co., is a very unusual bank variation that Banthrico produced in two sizes.

The Improvement Savings and the American Fletcher bank variations are all the same sizes.

The earlier variation of the Exchange National Bank is the shorter and a more difficult bank to locate (#580), while the later variation is taller and easier to find (#581).

The Savings Post comes is two different sizes and they are generic banks. The first version is taller (#690) than the second version (#691). They both are hard to find and rated "D" banks.

Hobart Federal Savings has two variations that are entirely different buildings. The first variation is the financial institution (#632) and the second version is "Old Hobart Mill" (#633) – a historical landmark in the area.

Two steel bank variations are pictured in #711A and #711B to show the color variations. These neat little steel buildings were first produced by Bankers Thrift Corp. and continued by Banthrico. They were probably first made in the late 1920s, in the 1930s, and the early 1940s (prior to WWII). They are still readily available.

There are three variations of the Northwestern National Bank (NW) tower. The first is engraved (#736), the second has a metal plate describing one of their banks (#737), and the third is the newest and is plastic (#739).

The Worcester Federal Savings has a variation. One (#745A) is a bank only and the other (#745B) is a bank and a pen holder on a marble base. The banks are the same size.

Another variation is #723 and #724. The first appears to be an earlier version of the Lincoln Memorial. Note that the steps, the height of the chair, the hands, and the hair are all different. The picture comes from the late Andy Moore collection and it is shown with hope that someone recognizes it and reports details to the author. The seco[nd] picture is the newer variation with the tin pop-ou[t] push-in trap that was produced in the 1970s.

A variation of the Arsenal appears (#532 a[nd] #533). The first is white metal and the second o[ne] is composition made in Japan for Banthrico. Th[ey] are the same height, but have different traps a[nd] years they were made.

The Miami Beach Federal Savings and Lo[an] Assn. has two very different variations. The fir[st] (#656) is a one-story building with a dome a[nd] was issued in the 1940s. The second (#657) is [a] high-rise building executed in 1957 with a nam[e] change.

The East Alton Savings & Loan building (#57[4]) is identical to First Federal Savings and Loan Ass[o]ciation of East Alton building (#593), except t[he] name and metal plate. It appears #574 is the fir[st] issue and #593 is the second issued since "Fe[d]eral" has been added to the bank name.

The B&O Queen City Station (#535) a[nd] Holmes County Court House (#634) are two [of] the last white metal banks that Banthrico produce[d]. In fact, Holmes County was completed the ye[ar] (1984) before their closing. See Chapter 14 f[or] new information on the B&O Queen City bank.

The Statue of Liberty (#727) also was ma[de] right before Banthrico sold the company. It appea[rs] that the Statue of Liberty was produced in 198[4-]1985 for Union Bank and Trust Co., Grand Ra[p]ids, Michigan. The banks had to be produced, di[s]tributed, and sold to help in the restoration of t[he] Statue of Liberty for the Centennial year of t[he] statue's dedication – "1886-1986." Chapter 1[4] has several Statue of Liberty pictures with box[es] that provide additional information on the ban[k]. See Chapter 14 for additional information.

Some historic and interesting banks appear. T[he] Beverly Savings (#544) (later became Beverly ban[k]) was the financial institution where Andy Moor[e], premier bank collector and author, served as t[he] President and Chairman of the Board. The First Fe[d]eral Savings of Chicago tower bank (#733A/ #733B) is an unusual bank. The pictures show t[he] front and back of the bank in the two color vari[a]tions. The banks had $1 pocket watches attache[d] with screws and they could be rotated and remove[d]. Some of the watches were "Westclox" and othe[rs] were "Ingersolls." They may have used other cloc[k] makers. The Exchange National Bank of Chica[go] (#579) is a rather unique Banthrico bank becau[se] the casting of the small cars (note the vintage) [is] separate from the rest of the bank. They had to b[e] attached by hand. The International Exposition [-] Travel and Transportation Building - Chicago 193[3] (#640) is an early (almost primitive) white met[al] bank by Banthrico. It has one of the earliest tra[ps] used by Banthrico. The bank was also manufa[c]tured by Preferred Bank Service Co. of N.[Y.] Banthrico and Rehberger may have cast some [of] these banks for Preferred Bank Service. Both com[]panies were in the city where the exposition w[as] held.

The majority of the buildings were built f[or] financial institutions, but seven buildings are insu[r]ance companies (#539, 565, 642, 644, 648, 65[9], 699), six are homes (#529A, 635, 638, 650, 70[2], 706), two are court houses (#634, 687), four a[re] historical buildings (#532, 633, 676, 710), tw[o] are printing plants (#523, 669), one is a steel pla[nt]

#684), one is a sports stadium (#622), one is an position hall (#640), one is a capitol building #546), one is a cache (#525), one is a railroad ation (#535), one is a Federal Government Building (#601), and one is an International Building #639). There are also office buildings that had ultiple usage such as the Sears Tower (#692) and e Wrigley Building (#721).

The Citizens National Bank in Chicago, Il. #556) was very hard to identify. Several bank and uilding collectors got involved in identifying the cation of the bank. We first thought it was the tizens National Bank of Freeport, N.Y., (discussed arlier) and Tom Hagan (Ohio) had the Freeport bank in his collection, but sold it. Tom informed me the bank he had (thought it was a Banthrico) was definitely marked with "Freeport N.Y." and the one I had (pictured) was not marked. Andy Anderson (Illinois) had the bank (#556) in his collection and I had taken a picture of it earlier in Chicago. Lance Oldham (Washington) wrote me that Michael Emrick of Tennessee (SBCS) had the Freeport bank, so I called him for a picture. He sent the same picture that we had of the Anderson bank. James Gass (Tennessee) sent me a picture of the Citizens National Bank (#556) and thought it was the Freeport bank, but had no idea of its location. Andy had purchased the bank in an antique mall in Chicago so he went back to the area to see if he could get further information on the bank. He talked with one of the antique dealers in the area who said that the picture of the Citizens National Bank (#556) that Andy had was indeed the Citizens National Bank that was located across the street. It had been torn down and replaced by another building. So, we feel fairly certain that the Citizens National Bank of Chicago has been properly identified. Thanks to all of these people for their assistance in finding the location of the bank. The work that went into locating the information on this bank is an example of the research that went into identifying many of the banks in this book.

23. Addressograph – Multigraph Corp.
½", T6, S, 1949, E
ome are Converted Banks
R.M. Ward 1949 Hundred Club"
ame in green woodgrain box

524. Alaska Mutual Savings Bank
(Anchorage, AK)
4 ½", T5, S, 1974, D

25. Alaska State Bank (AK)
5/8", T5, G, c. 1960s, C
uilding Cache

526. Alexandria National Bank
(Alexandria, VA)
3 3/8", T4, S, c. 1950s, E

527. Alliance Savings & Loan Assn.
(Chicago, IL)
2 ½", T4, S, 1949, C
"Saving Time: 12:36"
Clock on roof

528A. American Fletcher National Bank
& Trust Co. – Indianapolis, (IN)
4", T4, S, c. 1960s, C
Monument picture on back
Attached metal plate variation
"On the circle"

528B. American Fletcher National Bank
(Indianapolis, IN)
4", T6, S, c. 1970s, C
Engraved variation

529A. "The American Home"
First National Bank – Winnetka (WI)
2 ¾", T3, G, c. 1940s, C
Plastic house with steel bottom

529B. "The American Home"
Maplewood Bank & Trust Co.
2 ¾", T3, G, c. 1940s, A

529C. "The American Home"
2 ¾", T3, G, c. 1940s, B
"Save and Have"

29D. "The American Home"
ome Federal S&L – Marion, OH
¾", T3, G, c. 1940s, A

529E. "The American Home"
Home Federal S&L
2 ¾", T9, G, c. 1940s, A
Collinsville, (Il) -
"The home for your savings"

30. American National Bank &
ust Co. – Chattanooga, TN
½", T4, S, 1968, C
aper sticker on top of tower:
The New American National Bldg.
eptember 1968"

531. Anacostia National Bank
(Washington, D. C.)
2 ¼", T1, S, c. 1950s, D
"Established 1910" (2 sides)

32. Arsenal Savings and Loan
ssn. – (Indianapolis, In.)
1/8", T4, S, c. 1960s, D
The Arsenal - Established 1885"

533. The Arsenal – Composition
(Indianapolis, In.)
3 1/8", T11, S, c. 1980s, D
"Made in Japan for Banthrico Inc. -
Chicago, Ill."

534. Aurora National Bank
(Aurora, Il.)
5 5/8", T4, S, c. 1960s, C
"Lighting your way to security"
Large torch on back of bldg.
Slot in top - Woman on front with torch

535. B and O Queen City Station
Cumberland, Maryland
2 7/8", T8, S, 1986, C
"Chessie Federal C.U.
Est. here 1948"
Valley Casting bank, see Chapter 14 information

536. Banco Atlantida
Acaba, Mexico
3 ¼", T4, S, c. 1950s, F
"Quien Trabaja, Principia Bien:"
(Those who are industrious have a good principle)

537. Bangor Savings Bank (ME)
3 3/8", T10, S, c. 1980s, C
"1852 – 1912"
Hole in roof for bills

538. Bank of Madison
3 5/8", T4, S, c. 1960s, D
Slot in Top
Sign Outside: "Bank on Madison" -
Clock: time "9:27"
"The bank with the friendly atmosphere"

539. Bankers Life Company
Des Moines (IA)
3 3/8", T4, S, c. 1950s, C
"Established 1879 -
Erected 1939"

540. Bankers Savings & Loan Assn.
(Galveston, TX)
1 7/8", T6, S, c. 1960s, D
"BSL" over entrance

541. Bankers Trust Company
New York, N. Y.
6", T1, S, c. 1950s, D

542. Bayside National Bank
of New York – Bayside, N. Y.
2 3/8", T1, S, c. 1940s, D
"Member Federal Deposit
Insurance Corporation"
Name on roof of bldg.

543. Beckley National Bank
(Beckley, WV)
2 ½", T8, S, c. 1960s, D
Slot in top also
Name on both sides

544. Beverly Savings
Chicago 20, Illinois
4 3/8", T4, S, c. 1960s, D
"8001 South Western Avenue -
Phone 436-3500"
"BSL" on roof

545. Brookfield Federal Savings
(Brookfield, IL)
2 1/8", T4, S, c. 1970s, C

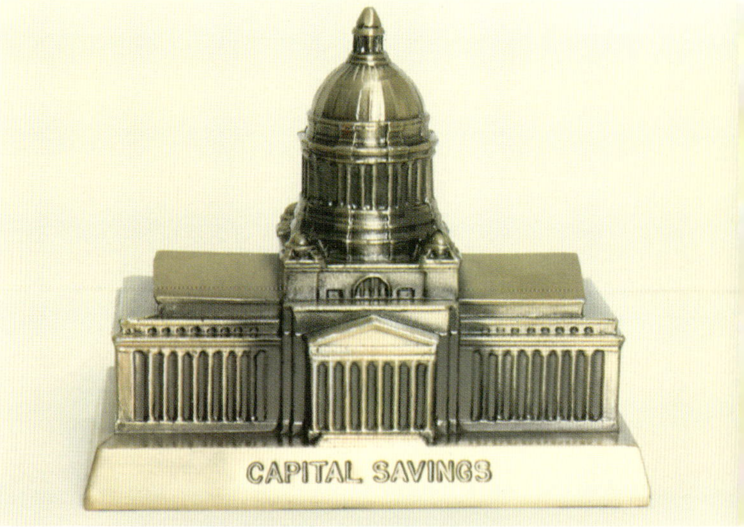

546. Capital Savings
Olympia, WA.
4 ¾", T8, G, 1978, B

547. Central Bank of Tampa (FL)
2 ¼", T1, S, c. 1940s, D
"Auto Teller Windows"
"Grand Central at Howard - Safe Deposit Vault"

548. Chelsea Savings Bank
(Belle Plaine, IA)
3 ¼", T4, S, 1954, D
"1854 - Our 100th Year - 1954"
Slot in top

549. Chemung Canal Trust Co. (NY, NY)
4", T8, S, 1983, C
"1833 – 150 Years of Banking – 1983"

550. Chesterton Rural Loan
& Savings Assn. (IN)
3", T8, S, c. 1970s, D

51. Chicago Federal Savings (IL)
3 ¾", T4, S, c. 1950s, D
lot in Top
"State and Washington - Downtown
emperature (with gauge)"
Save for Better Living"

552. Chittenden Trust Company
Burlington, Vermont
2 7/8", T8, S, 1971, D

53. Citizens Federal Savings
leveland, (Ohio)
5/8", T9, S, c. 1970s, C
riangular bldg.

554. Citizens Federal Savings and
Loan Assn. Hammond, Indiana
3 ¾", T4, S, c. 1950s, D
Name also on sign projecting
above building

55. Citizens National Bank
bilene, Texas
1/8", T4, S, c. 1950s, D
ame and city on 2 sides

556. Citizens National Bank
(Chicago, IL)
3 ¼", T8, S, c. 1970s, E

557. Citizens Savings and Loan Co.
Painesville, Ohio
2 ¾", T4, S, c. 1960, C
"Serving safely since 1889"
Slot in top

558. The City National Bank
of Houston (TX)
5 ½", T4, S, c. 1950s, C
Name and city on 2 Sides
Slot in top

559. Clearing Bank
(Chicago, IL)
2 1/8", T9, S, c. 1970s, B
"Bank from your car - 4 drive-in windows" - Same building as State Bank of Clearing #702

560. Clyde Savings
(North Riverside, IL)
3 5/8", T6, S, 1971, A
Name on 2 sides

561. Coast Federal Savings and Loan Assn. – Los Angeles, (CA)
5 1/8", T3, S, 1961, E
Slot in top

562. Colorado Federal Savings
Denver (CO)
2 3/8", T4, S, 1959, D
"Colorado Fed S&L Assn".
Mountains on back of bank/slot in top

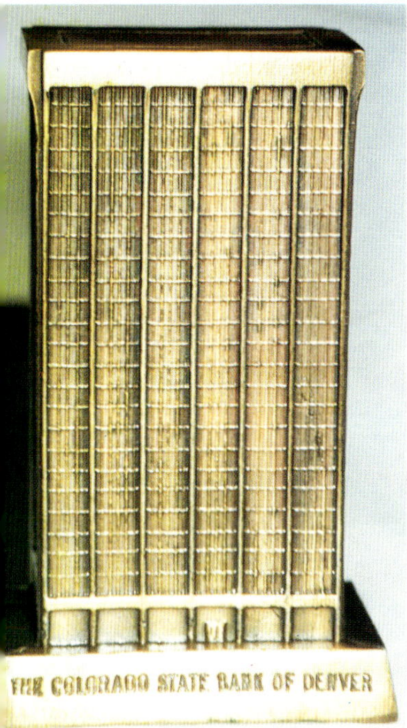

563. Colorado State Bank of Denver (CO).
5 5/8", T5, S, 1971, D

564. Commonwealth National Bank of San Francisco (CA)
3 ¼", T4, S, c. 1960s, D
Name on all sides
Slot in top

565. Continental Assurance Company (Chicago, Il.) (LEAD)
5", T2, S, c. 1930s, E
Soldered corners on base plate.
One of earliest Banthrico buildings.

566. Cook County Federal Savings – (Chicago, IL)
4 ¼", T4, S, 1954, C
Name over main entrance
"Authentic replica of an historic Williamsburg (VA) building and an exact copy of our new home"
(per Banthrico)

;7. Dade Federal Savings
'ade County, FL.)
'5/8", T4, S, c. 1960s, D
One of the nation's oldest and
'gest - Flagler at First"
ɔt in top also

568. DeLand Federal Savings (FL)
3 1/8", T4, S, 1956, C
Bank Name on Both Sides/Roof Sign
"Savings Insured" (2 sides)

569. Des Moines Savings and Loan Assn – (Des Moines, IA)
4 3/8", T3, S, c. 1940s, D
"The Oldest in Des Moines" - "210 Sixth Avenue" (on roof)

570. The Dollar Savings Bank Pittsburgh, PA.
3 1/8", T4, S, c. 1940s, C
"Organized in 1855 - 340 Fourth Ave.
The only mutual savings bank in Pittsburgh"

571. Dover Co-operative Bank (Dover, DE)
2 ½", T4, S, c. 1950s, C
Name on 2 sides - Slot in top

572. Drive-In Bank with Car
4", T19, G, 1978, D
Plastic Mechanical
Merchants Nat Bk of Aurora (Il).
"Made in British Hong Kong for Banthrico, Inc. - Chicago"

573. Durham Life Insurance Co. (Durham, N. C.)
4 ½", T4, S, c. 1960s, D
"We protect the family"
Name and motto on 2 sides

574. East Alton Savings & Loan – (Alton, IL.)
2 1/8", T4, S, c. 1960s, C

575. Englewood State Bank
Englewood, Colorado
3 1/8", T4, S, c. 1950s, D
Name on roof sign - both sides.
Blue and white painted sign.

576. Equitable Building
Chicago, Illinois
5 1/8", T4, S, 1965, D

577. Erie County Savings Bank
Buffalo, New York
6 1/8", T4, S, 1969, C
"One Main Place - Big E"
Slot in top of tower

578. The
Exchange Bank of
St. Augustine (FL)
5 ½", T4, S, c.
1950s, D
Name on 2 sides.
Slot in roof.
Note 2 doors on
front for
vault entrances

579. The Exchange National
Bank of Chicago, (IL)
½", T1, S, c. 1950s, E
"Originators of complete
autobanking" on roof. Note 4 cars
leaving drive-in exit

580. The Exchange National
Bank
Colorado Springs, (Co.) (Earlier)
3 ¾", T1, S, c. 1956, E
Slot and currency hole on top.
Some lead versions.

581. Exchange National Bank
Colorado Springs (CO) (Later)
4 ¼", T4, S, c. 1960s, C
Bank name sign on
corner of bldg. - Slot in top.

582. Fall River Trust
Co. (Fall River)
& Somerset (MA)
4 5/8", T4, S, c.
1950s, D
"Main Office - 43
N. Main St.
Branches: South End
- East End -
Somerset" Slot in
top.

583. Farmers & Mechanics Savings Bank
Framingham, Mass
2 5/8", T4, S, c. 1960s, D
Slot in roof

584. Fayette National Bank and
Trust Co. – (Fayette, Il.)
4 ¾", T4, S, c. 1960s, E
Slot in top oval

585. Fidelity Bank & Trust Co.
Indianapolis, (IN)
4", T4, S, c. 1960s, D
Predecessor of American
Fletcher Nat Bk & Tr. Co.

586. Fidelity Federal Saving and
Loan Assn. – Greenville, S. C.
2 ½", T4, S, c. 1960s, D
"Since 1921" - Slot in roof

7. First Federal Savings
ıambra, (CA)
¾", T4, S, c. 1960s, D
tate Street - Garfield Ave"
me on 2 sides of roof sign

588. First Federal Savings
of Berwyn, (IL)
2 7/8", T1, S, c. 1940s, D
Founded 1927-Stanley Ave at Oak Park"
Clock projecting on front of building

9. First Federal Savings
Porte, Indiana
5/8", T4, S, c. 1960s, D
st Federal" on tower
sides)

590. First Federal Savings
Raleigh, N. C.
4 1/8", T6, S, c. 1960s, E
Slot in roof.

591. First Federal
Savings
St. Petersburg, Florida
6", T4, S, c. 1950s, D
"U.S. Charter No. 3 -
One of
America's largest"
"First Federal (four
sides)

592A. First Federal Savings and
Loan Assn – Detroit, Michigan
3", T1, S, c. 1930s, E (LEAD)
"Griswold at Lafayette"
"Accounts insured up to $5,000".
Base plate soldered at corners.
Early version.

592B. First Federal Savings and
Loan Assn. – Detroit, Michigan
3", T1, S, c. 1940s, D (WM)
"Accounts insured up to $10,000"
Brads on base plate corners

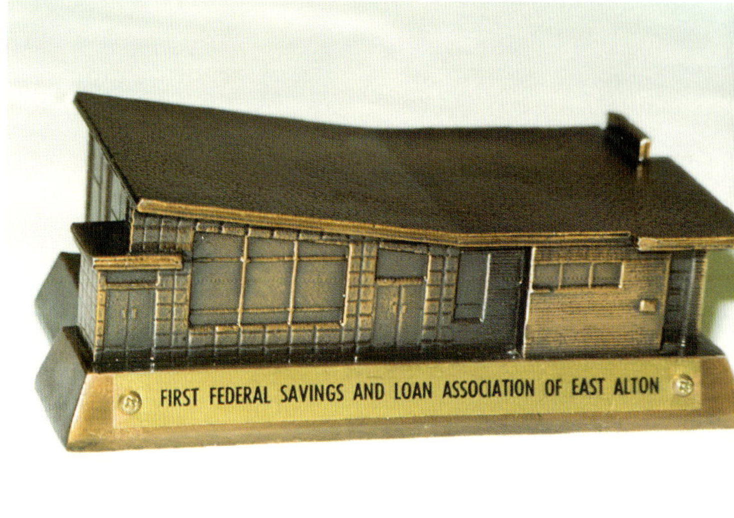

593. First Federal Savings and
Loan Assn. Of East Alton, (Il)
2 1/8", T1, S, c. 1940s, C
Same building as #574
Metal name plate variation

594. First Federal Savings and
Loan Assn of Elgin, (Il).
1 7/8", T5, S, c. 1960s, C

595. First Federal Savings and
Loan Assn. – Greenville, S. C.
2 ¼", T4, S, c. 1960s, C

596. First Federal Savings and
Loan Assn. – New Castle, Pennsylvania
3 ¼", T4, S, c. 1960s, D
"First Federal Plaza

597. First Federal Savings and
Loan Assn. – Puerto Rico
3 ¾", T4, S, c. 1950s, F
"Federal Savings" on sign.
"Parada - 22 1/2" -
Slot in roof

8. First Federal Savings and
an Assn. – Wade Hampton, (Il)
/8", T4, S, c. 1960s, D
Main Office: 320 Buncombe St."

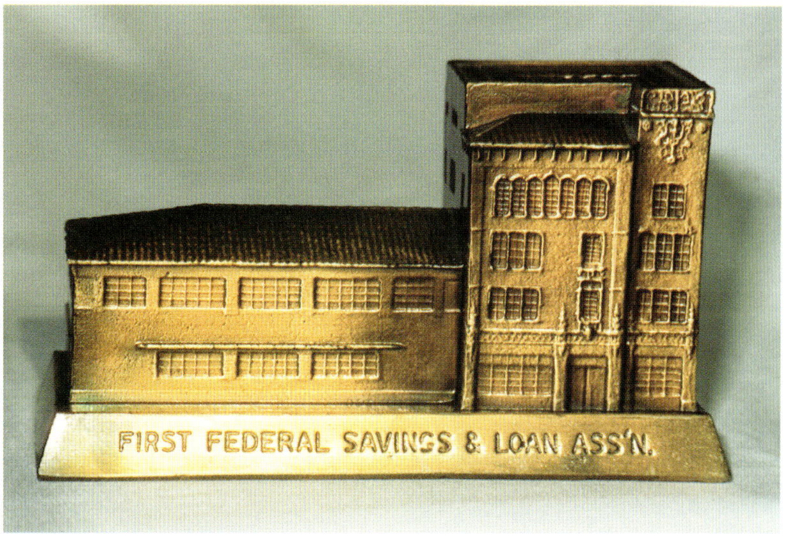

599. First Federal Savings and
Loan Assn. – West Palm Beach, FL
3 ½", T4, S, c. 1950s, D
Slot in top

0. The First Liberty National Bank
erty, Tx.)
7/8", T4, S, c. 1960s, E
t in top

601. First and Merchants National
Bank – (Arlington, VA) Pentagon
1 5/8", T4, S, c. 1960s, D
"Your Pentagon Bank"
Slot in roof

2. First National Bank
narillo, Texas
, T4, S, c. 1960s, C
t in roof

603. The First National Bank
of Atlanta, (GA)
5 ¾", T5, S, 1968, D

604. First National Bank
of Casper, (WY)
4", T4, S, 1957, E
"Save First"

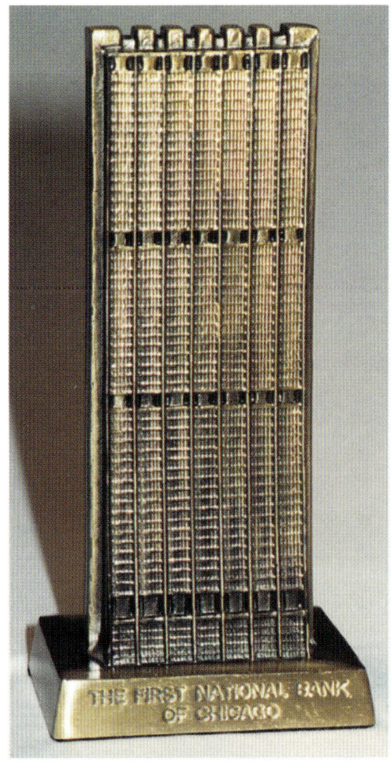

605. The First National Bank
of Chicago, (IL)
7 5/8", T9, S, 1969, B
1965-1969 - Construction dates.
Architect C.F. Murphy

606. First National Bank
Conroe, Texas
2 ¼", T4, S, c. 1950s, D
"Since 1914" - Bank name, city
and date on 2 sides
Slot in top.

607. First National Bank
of Danville, (IL)
3", T1, S, c. 1930s, E (LEAD)
Name on 2 sides and above front
door. Early Banthrico. Corners on
base plate soldered - Great detail

608. First National Bank
of Denver, (Co.)
5 5/8", T4, S, c. 1960s, D
"Providing Banking Service Since
1860" "First National Bank" on
both sides of tower

609. First National Bank
of Des Plaines, (IL)
3 3/8", T9, S, c. 1970s, C

610. First National Bank
Jackson, Mississippi
5", T4, S, c. 1960s, D
"First National Bank" (2 sides)

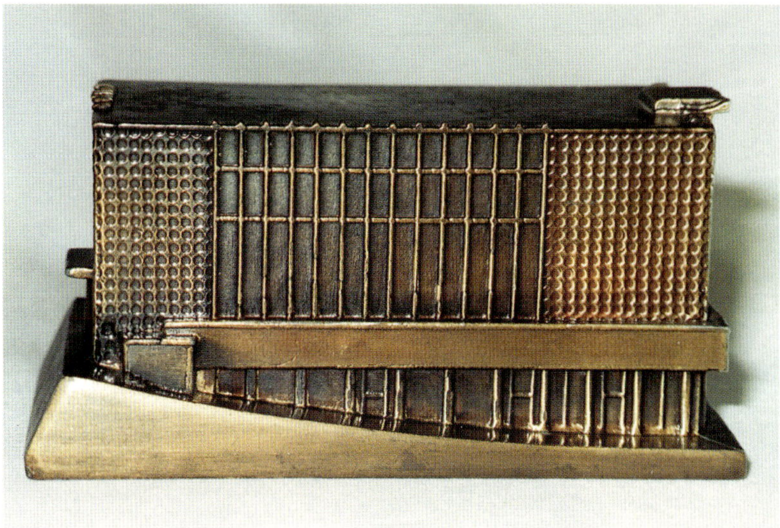

611. First National Bank
Montevideo, Minn.
3 1/8", T4, S, c. 1960s, D
Name on back of building

12. First National Bank
pringfield, Ill.
5/8", T8, S, c. 1970s, C

613. First National Bank
of Sterling, Illinois
2 3/8", T4, S, 1966, D

14. First National Bank (USA)
¼", T4, S, c. 1960s, E
"1st" - round medallion on
de of building

615. First National Bank
Wamego, Kansas
2 ½", T9, S, c. 1970s, C

616. First National Bank
of Warren, (RI)
4", T4, S, c. 1960s, D
Name on 2 sides of building
Slot in top

617. The First National
Bank
and Trust Co. – Tulsa,
(OK)
5 ½", T4, S, c. 1950s, D
"1st of Tulsa" (2 sides)
"Think first - pay yourself
first"

618. The First New Haven
National Bank (CT)
5 1/8", T4, S, 1961, C
Bank name (2 sides of tower)

619. First Security Bank
Salt Lake City, (UT)
5 ½", T4, S, 1956, C
"First Security" (2 sides of tower)
"The bank for everybody"

620. First Security Trust and
Savings Bank – (Chicago, Il.)
2 1/8", T4, S, c. 1960s, D
"Drive-In Windows - Free Parking"

621. First State Bank (USA)
2 5/8", T9, S, c. 1970s, D
"Est. June 4, 1934"

22. First State Bank of Bellaire -
ouston, Tx.) (Astrodome)
1/8", T4, S, c. 1960s, C
arris County Domed Stadium
ome of the Houston Astros
seball team (National League)

623. Ford City Bank
(Chicago, Il.)
3 ¾", T9, S, c. 1970s, C
Note two walk through areas
in building. Unusual.

624. Fort Wayne National
Bank – (Fort Wayne, In.)
5 3/8", T9, S, c. 1970s, B

625. Gate City Savings
Fargo, (N. D.)
4", T4, S, c. 1960s, D
"Fargo's Saving Center"

ft: 626. Gilbraltar Savings and
ilding Assn. – Houston, Texas
7/8", T4, S, c. 1950s, C
Where safety and profit meet -
braltar 1921"
ot in roof.

ght: 627. Goldome (Buffalo
vings Bank)
uffalo, N. Y.)
5/8", T8, S, c. 1970s, C
Formerly Buffalo Savings Bank -
ilt March 1901 -
chitect - Edward B. Green"
imited Edition 1 - #3698"

628. The Great Northern Building and Loan Co. – (Barberton, OH)
2", T4, S, c. 1950s, C

629. Hattiesburg Federal Savings – (Hattiesburg, Ms.)
3 1/8", T5, S, c. 1960s, D
Name on Both Sides - Slot in roof

630. Heights State Bank (Houston, TX)
2", T4, S, c. 1960s, D
Slot in top

631. Heritage County Bank (Blue Island, IL)
3 ½", T9, S, 1980, C
"1930 - 1980"

632. Hobart Federal Savings (Hobart, In.)
2", T4, S, c. 1960s, D
"Savings" on top of bldg. edge. Slot in roof. "Hobart Federal Savings" on sides/end of bldg.

633. Hobart Federal Savings (Hobart, IN.)
"Old Hobart Mill"
4 5/8", T6, S, c. 1960s, D

34. Holmes County Courthouse
Millersburg, Ohio)
¼", T6, S, 1984, C
"1884-1984 - 100 Years"
1879 - The Commercial and Savings
Bank-1984 - 105 Years". One of last
uilding banks made by Banthrico.

635. House with Dormers
2 7/8", T7, G, c. 1940s, B
Plastic
Trap has hole for currency as well as slot.

36. Hoyne Savings and
oan Assn. – (Chicago, Il.)
¼", T4, S, c. 1960s, D
"Safe Since 1887 - Milwaukee at
awrence". Slot in top

637A. Improvement Savings and
Loan Assn. – (Aurora, Il.)
2 ¼", T4, S, c. 1950s, C
"Galena Blvd. - Oak Ave." Name on roof

37B. "Now Improvement Federal"
Aurora, IL)
¼", T4, S, c. 1960s, C
Name painted on back. Old name
still on roof.

637C. Improvement Federal Savings
and Loan Assn. – Aurora, Illinois
2 ¼", T4, S, c. 1960s, B
"Federal" added to name

638. Industrial Bank
2 5/8", T3, G, c. 1940s, B
Plastic/Metal Name Plate
"2% on savings"
steel bottom

639. International Building, Credit
Union Center – Madison, Wisconsin
2 3/8", T9, S, c. 1970s, C
"U.S.A." in title

640. International Exposition - Travel
& Trans. Bldg. – Chicago 1934
3 1/8", T20, S, 1934, D
"A Century of Progress" Some have 1933
date. Made in copper & silver finishes

641. Iowa – Des Moines National
Bank
(Des Moines, IA)
5 ½", T8, S, c. 1970s, C

642. Jefferson Standard
Life Ins.
Co. – (Greensboro, N. C.
)
5 5/8", T4, S, 1957, E
"Fiftieth Anniversary -
1907-1957"
Rehberger also did bank
with name. No name on
this bank.

643. Joliet Federal Savings
(Joliet, Il.)
2 ¼", T4, S, c. 1950s, C
"Will County's Largest Savings
Institution"

644. Lamar Life Building
Jackson, Mississippi
6", T8, S, 1959, C
"Home Office, Lanar Life Ins. Co. - "The Pride of the South" Clock in tower (all sides)

645. Lansing Savings and Loan Assn. – (Lansing, Il.)
2", T5, S, c. 1960s, C

646. Lawn Savings
(Evergreen Park, Il)
3 1/8", T4, S, c. 1960s, B

647. Leon Federal Savings and Loan Assn – Tallahassee, Florida
4 3/8", T4, S, c. 1960s, D
Eagle on back. Slot in top also.

648. Life of Georgia
(Atlanta, GA)
6 5/8", T4, S, c. 1960s, D
"Life of Georgia" (4 sides) Ins. Co.

649. Livestock National Bank
Chicago, (IL)
6", T1, S, c. 1940s, D
"4150 So. Halstead St."

650. Greater Louisville First Federal
S&L Assn. – Louisville, (KY)
2 5/8", T3, G, c. 1940s, B
Plastic house - metal name
plates.

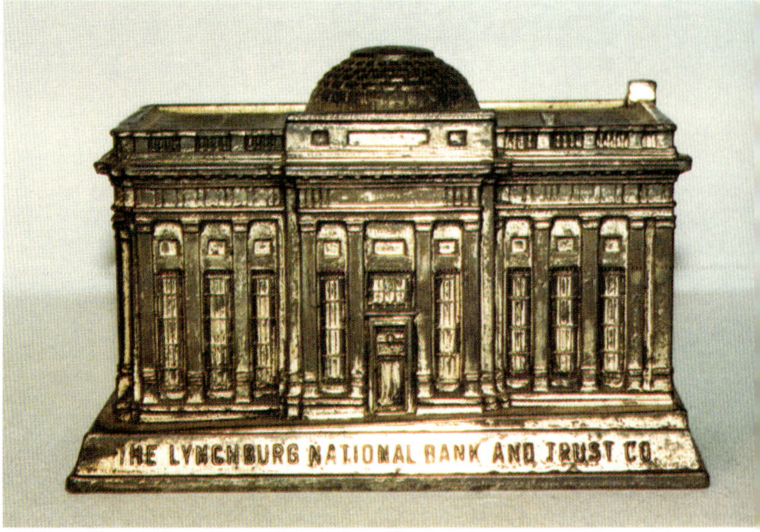

651. The Lynchburg National Bank
and Trust Co. – (Lynchburg, VA.)
3 1/8", T1, S, c. 1930s, E (LEAD)
Name of bank (2 Sides). Bill hole in trap

652. Marconi Savings and Loan
Assn. – (Philadelphia, PA.)
2 3/8", T4, S, c. 1950s, D

653. Marion County State Bank
(Knoxville, IA?) (LEAD)
2 5/8", T1, S, c. 1930s, F
"Franklin St. - Main St." Bill hole
in bottom. Early Banthrico bldg.

654. Merchants Bank
Allentown, PA
2 5/8", T5, S, c. 1960s, C
Slot in top also

655. Merchants National
Bank and Trust Co. –
Indianapolis, (IN)
4 1/8", T1, S, c. 1940s, D
"The Oldest National Bk &
Trust Co. in Indianapolis-
Established 1865" "An
office in your neighbor-
hood" Slot in top also. State
of In. in relief behind name.

56. Miami Beach Federal Savings
Loan Assn – Miami Beach, (FL)
¾", T1, S, c. 1940s, E (EARLIER)
Save-A Svgs Institution"-1st Version

657. Miami Beach
Federal Savings
Miami Beach, (FL)
(LATER)
5 ½", T4, S, 1957, C
"Lincoln Road-
Washington Ave."
Name on back.
Second version

58. Mid Kansas Federal Savings
Vichita, KS)
1/8", T4, S, c. 1960s, C
'Center of Savings-230 S. Market"
lot in top

659. Mid-West National
Insurance Companies (TN)
3 ¾", T4, S, c. 1960, B
"MWN" on globe/map of US on
top of roof.

60. Midland Savings and Loan
ssn. – Midland, Texas
", T4, S, c. 1950s, D
lot in top

661. Minot Federal Savings
Minot (N. D.)
2 ½", T4, S, c. 1950s, E
"Time - 7:54 - Temperature" sign

662. The Moffat County State Bank
Craig, Colorado
2", T4, S, c. 1960s, D
Car going in bldg. and one coming out.
Fireplace at one end of building.

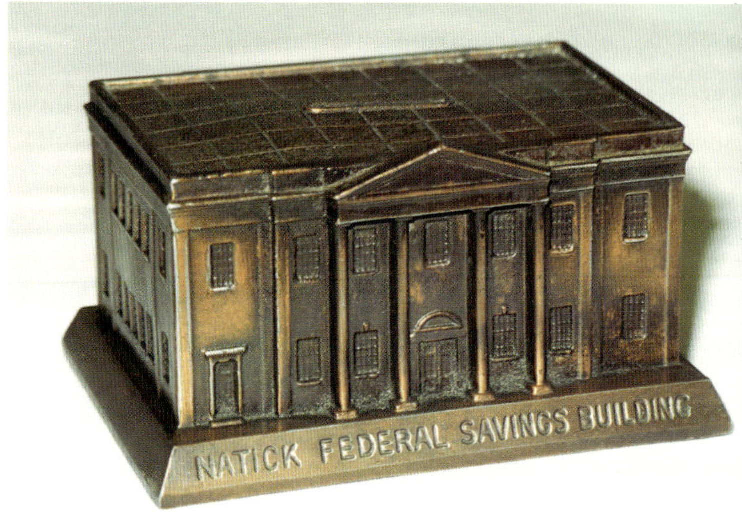

663. Natick Federal Savings
Building – (Natick, MA.)
2 7/8", T4, S, c. 1940s, E (LEAD)
Slot in roof.

664. National Bank of Albany
Park (Chicago, Il.)
2", T5, S, c. 1960s, C

665. The National Bank of Commerce
Jackson, Tenn.
2 7/8", T1, S, c. 1950s, D
Name and city on 2 sides.
Slot also in top.

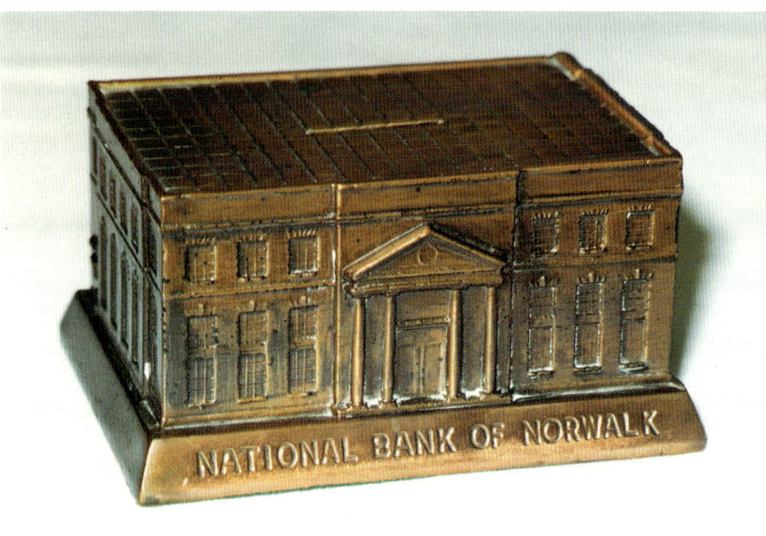

666. National Bank of Norwalk
(Norwalk, CT.)
2 ½", T4, S, c. 1950s, C
Slot in roof.

667. The National City Bank
of Evansville – Evansville, Indiana
2 ½", T1, S, c. 1930s, E (LEAD)
"Bank the City Way - Third and Main"
Bank name and additional info. on roof.

68. New Britain Federal Savings &
oan Assn. – New Britain, Connecticut
1/8", T4, S, c. 1960s, C
"Serving Since 1886
4 West Main St." Slot in top.

669. New Method Book Bindery
Inc. (USA) (CONVERTED)
1 ¼", T6, S, c. 1960s, D
"Bound to Stay Bound - Books"
over entrance. Name on roof.

70. Newport Balboa Savings
nd Loan Assn. (CA)
3/8", T4, S, c. 1950s, C
lot in top.

671. The Northern Trust Company
Bank – (Chicago, Il.)
4", T4, S, 1961, C
"Chicago F16-5500 - Principles
that Endure - LaSalle and Monroe"
Anchor emblem at end of building.
Slot in roof.

672A. Northwest Federal
(Spencer, Ia.)
1 ¾", T4, S, c. 1960s, C

672B. Northwest Federal
Limited Edition/Gold Plated
Note Raised Lettering - D

673. Norwood Savings and
Loan Assn. – (Chicago, Il.)
2", T4, S, 1972, C

674. Nutley Savings
(NJ)
2 ¼", T5, S, c. 1970s, C

675. Oak Lawn Federal
Savings – (Chicago, Il.)
2 ¾", T4, S, c. 1950s, C
"5555 West 95th St."

676. Old Capital Bank and
Trust Co. – Corydon, Indiana
4 ¼", T8, S, c. 1970s, C
"1816 - Corydon-1825"
"Indiana's First Capitol"

677. Old Second National
Bank – Aurora, Illinois
3 ¾", T4, S, c. 1950s, C
"Established 1871 - Drive-Up -
River Street"

678. Olympic Savings
Berwyn, Ill.
3 ½", T4, S, c. 1960s, C
"Look to the Olympic Torch for
Security - 6201 Cermak Rd. Torch
on building roof. Time: 12:26

9. Orlando Federal Savings
lando, (FL)
/8", T4, S, c. 1950s, D

680. Petersburg Mutual Savings
Petersburg, Virginia
4 5/8", T4, S, c. 1950s, D
Name on 2 sides of roof top.

1. Petersburg Savings and American
ıst Co. – (Petersburg, Va.)
/4", T4, S, c. 1950s, D
The Oldest Bank in the Oldest State
the United States - 1860" (2 sides)
t in top

682. Port Angeles Savings
& Loan Assn. (WA)
2 3/8", T4, S, 1954, D
"101 West Front St. - Washington"
Slot in top.

3. Pullman Bank and
ıst Co. – (Chicago, Il)
5/8", T8, S, 1974, B
Heritage Bank"

684. Ralston Steel Corporation
Skokie, Illinois
1 5/8", T4, S, c. 1960s, D
"Save with Ralston - Everything in
Steel". Vat pouring steel on rooftop

685. Rapides Bank and Trust
Co. – Alexandria, Louisiana
2 5/8", T4, S, 1965, F
Slot in roof.

686. Red River National Bank
(Shreveport, L. A. ?)
2 5/8", T4, S, c. 1950s, E
"Save Today for Tomorrow -
Thrift Pays". Location: Bossier
City, La. ?

687. Rutherford County Courthouse
Tennessee
5 ¼", T9, S, 1976, C
"Erected 1859 - Cast 1976"
(bank)

688. Salem Federal Saving
& Loan Assn. – (Salem, OR.)
1 ¾", T4, S, 1961, D

689. Santa Rosa Savings and
Loan Assn. – (Santa Rosa, CA.)
2", T4, S, 1960, C
"Founded 1888"

690. The Savings Post
Western Building - Large
3 7/8", T1, G, 1952, D
"Hitch Your Future to the Savings
Post - Pat. 1-15-52
G. Flexner"

691. The Saving Post
Western Building – Small
3 5/8", T33, G, c. 1952, D
Same patent date/person

692. Sears Tower
(Chicago, IL)
7 ½", T9, S, 1974, B
"Home of Sears Bank and Trust Co."

3. Second National Bank
uston, Texas
7/8", T4, S, c. 1950s, D
t in roof.

694. Second National Bank
Nashua, N. H. (LEAD)
2 3/8", T4, S, c. 1930s, E
Name of bank on 3 sides.
Slot and currency hole in roof. Slot in bottom also.

5. The Security Bank & Trust
. – Lawton, Oklahoma
1/8", T1, S, c. 1950s, D
row pointing to "Drive-In Tellers"
ecurity Bank" over front door. Trap
t in upside down on this bank.

696. Security National Bank
Joplin, Missouri
1 7/8", T4, S, c. 1950s, D
"F.D.I.C." - Both ends

697. Serial Federal Savings and
Loan Assn. – New York City, (N. Y.)
2 1/8", T1, S, c. 1930s, E (LEAD)
"Established 1885. To Church St. at
Vessey St." Very heavy lead bank.

698. South Main State Bank
Houston, Texas
2 ¾", T1, S, c. 1940s, D
"Your Doorway to Thrift - Every Banking
Facility" Slot in roof.

699. Southern Republic Life
Insurance Co. – Houston, Texas
3 5/8", T1, S, c. 1950s, C
"Mighty Oaks from Little Acorns"

700. Southwestern Savings
& Loan Assn. – Houston, (TX)
3 7/8", T1, S, c. 1950s, E
"Liberal Dividends - 3401 Main" "Place
Your Savings Here" (slot in roof)

701. St. Paul Federal
Savings of Chicago, (IL)
2 ¾", T1, S, c. 1950s, B
"6720 W. North Ave."
Slot in roof

702. State Bank of Clearing
(Chicago, Il.)
2 1/8", T4, S, 1958, B
"Bank from your car - 4 Drive-In Windows"

)3. State Loan and Savings
ssn. – Granite City, Illinois
¾", T4, G, c. 1960s, A
ouse with car in garage.
304 Niedringhaus Ave. Each account
sured for $10,000". Info on Trap.
)ream House" bank

704. Staten Island Savings
Bank – (Staten Island, N. Y.)
1 ¾", T4, S, c. 1950s, D
Name and slot on roof.

)5. Steel City Federal
avings – Gary, Indiana
1/8", T4, S, c. 1960s, D
mbossed bucket of molten metal being
oured on back of bldg. "Savings
sured - 846 Broadway" Slot in top

706. Small Steel House
3 ¼", T3, G, c. 1930s, C
Bankers Thrift Co.
Appears to be an early Banthrico

707. Syracuse Savings Bank
(Syracuse, N. Y.)
5 5/8", T8, S, c. 1960s, D
"Oldest Bank in Syracuse - Established
in 1849"

708. Tallahassee Federal
Savings – (Tallahassee, FL.)
2 ¾", T8, S, 1959, C

709. Texas Bank
Dallas, (TX)
5 1/8", T4, S, c. 1960s, D
"America's Most Modern Auto-Bank. Time to Know You" Slot in top. Cars on roof and in garage.

710. Trout Hall – 1770
Allentown, PA
3 7/8", T8, S, 1979, C
"Ltd. Edition"

711A. Union Savings Bank
St. Charles, Mo. (STEEL)
2 5/8", T3, G, c. 1930s, C
"Save for a Home"
"B.T. Corp - Chgo" on trap

711B. State Bank of De Pere
West De Pere, Wis. (STEEL)
2 5/8", T3, G, c. 1930s, C

712. United Savings & Loan Assn. – (Chicago, IL.)
2 ¼", T4, S, c. 1960s, C
"Where You Save - Does Make A Difference" "4730 W. 79th St."

713. University State Bank
Houston, Texas
2 ¼", T4, S, c. 1950s, C
"Free Parking - Drive-In Windows".
Slot in top

714. Virginia Mutual Savings (VA)
4", T8, S, c. 1970s, E

715. Waltham Federal
Waltham, MASS.
2 ½", T4, S, c. 1956, C
"It Pays to Save at Waltham Federal - Founded 1880"

6. The Waterbury National Bank
aterbury, Connecticut
¾", T1, S, c. 1950s, C
Established 1848 - Erected 1921"

717. Wichita Federal Saving & Loan Assn. – (Wichita, KS)
2 7/8", T4, S, 1955, C
"Witchita Federal" sign on roof.
"Broadway at Waterman.
AM-5-3151"

718. Wilmington Mutual Savings & Loan Assn. – Wilmington, (CA)
2 ½", T4, S, c. 1950s, D
"1000 Avalon Blvd" Name on 2 sides. Slot in roof.

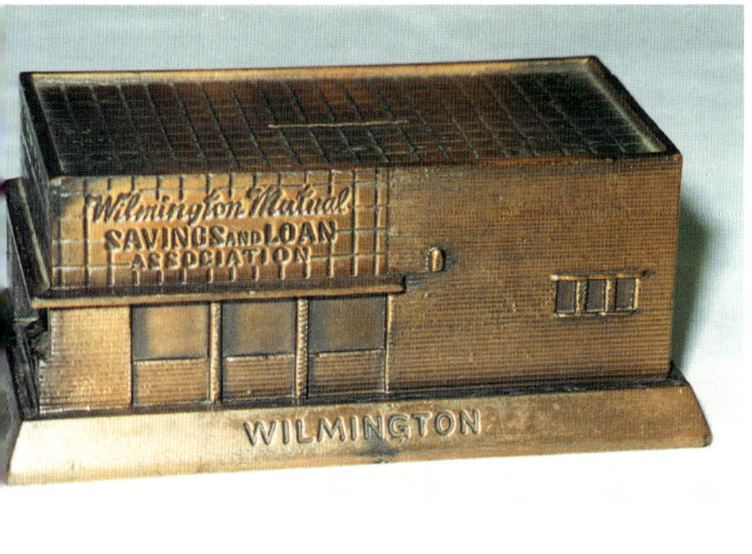

719. Winnetka Trust & Savings Bank – (Winnetka, WI.)
3 1/8", T4, S, c. 1960s, C
Car going in building and another coming out.

720. Worcester Federal Savings
Springfield, (MA)
3 5/8", T4, S, c. 1965, C
"New England's Largest - Main and Sanford"

721. Wrigley Building/Boulevard Bank – (Chicago, IL)
7 ½", T10, S, c. 1980s, C
Andy Moore's Bank (CEO)
White enamel paint

Left: 722. Buffalo Bill on Base
5 ½", T4, S, c. 1964, C
"July 4, 1924 – Cody, Wyoming"
"MDCCCXLVI - MCMXVII"
"The Scout - Col. William F. Cody"
"Sculptured and Presented by Gertrude Vanderbilt Whitney"
" ©1964 J.B. Minter"

Right: 723. Lincoln Memorial (OLDER)
(Washington, D. C.)
6 ¼", T1, G, c. 1940s, D
Older variation. Greater details. Note differences.

724. Lincoln Memorial (NEWER)
(Washington, D. C.)
6 ¼", T9, G, 1976, B
Newer variation.

725. The Old Prospector
Denver, Colorado
6 ¼", T4, S, 1963, C
Central Bank & Tr. Co. "A Landmark Atop the Mining Exchange Building - 1892-1963" Pick in one hand, ore in other

Opposite page; top left to right:
726. Oregon Pioneer
Salem, Oregon
7 7/8", T3, S, c. 1940s, D
The Commercial Bank. Statue on Oregon Capitol Bldg Axe in one hand, shelter cover in other. Slot in back at of base.

727. Statue of Liberty
(New York City, N. Y.)
7 ¾", T10, S, 1986, C
Union Bank & Tr. Co. - Grand Rapids, Mi. Union Bank contributed $2.00 for statue restoration for every ba sold. Banthrico did this bank in 1984-85 Prior to the closure. Sale of banks before Centennial in 1986 need to help in restoration of statue. "Liberty-Enlightening t World" Bank started by Banthrico and completed by V Casting.

728. Statue of Minuteman
(Concord, MA.)
8 1/8", T3, G, 1941, D
Standing by plow with rifle in hand and coat on plow. Symbo the American Revolutionary Soldiers ready at a minute's noti fight for their country. The following is on the statue base: B rude bridge that arched the flood. Their flags to April's breez unfurled Here once the embattled farmers stood And fired shot heard round the world.

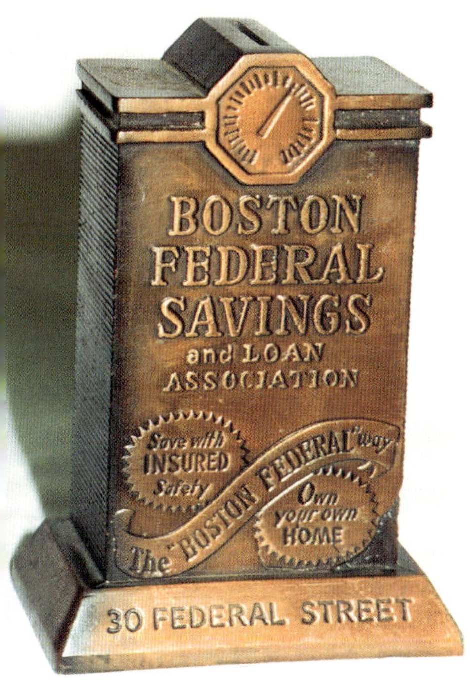

29. Boston Federal Savings & Loan Assn. – (Boston,
IA.)
1/8", T4, S, c. 1950s, C
"Save with Insured Safety - Where you save does make difference - 30 Federal Street". Temperature gauge in ont/back. Slot in top.

730. Citizens Bank
6 ¾", T4, S, c. 1960s, C
"CB" Tower with Ball
Temperature Code - Red: Warmer, Blue: Colder, Yellow: No Change, Flashing: Rain or Snow"

731. Citizens National Bank
Waco, Texas
5 5/8", T3, S, c. 1940s, D
"Citizen Time - 12:34 - Temperature" (2 sides)

732. Eastern Trust & Banking
Co. – Bangor, Maine
6 ½",T3, S, 1962, D
"Established 1887 - 75th Anniversary -
Branches: Old Town, Machias. Slot in top

733A. First Federal Savings
of Chicago, (IL) (FRONT)
7 1/8", T1, S, c. 1950s, D
Silver color

733B. First Federal Savings
of Chicago, (IL) (BACK)
7 1/8", T4, S, c. 1950s, D
Copper color

734. Key Biscayne Savings
& Loan – (Key Biscayne, Fl.)
7", T8, S, c. 1970s, D
Lighthouse tower

735. Mt. Greylock War
Memorial – (Lanesborough, MA.)
6 ¾", T8, S, c. 1970s, B

736. Northwestern National ("NW")
Bank – (Montevideo, Mn.)
6 7/8", T3, S, c. 1950s, C
Weatherball tower

37. Fifth Northwestern National
ank – (Montevideo, Mn.)
7/8", T3, S, c. 1950s, C
etal name plate variation

738. Northwestern National
Bank – Side View
Weatherball Code

739. Northwestern National
6 7/8", T31, S, c. 1960s, B
Bank – Plastic/Code
Slot in back

40. Northwestern National ("NW")
ank – (Montevideo, Mn.)
5/8", T4, S, c. 1960s, C
May We Help You Today?

741. Peoples Trust & Savings
Co. (Fort Wayne, IN ?)
5 3/8", T1, S, 1953, D
"50th Anniversary - 1903-1953 -
"The Bank Under the Clock"
Chesterton, In?

742. Riverside National Bank
(Riverside, IL.)
6 ½", T8, S, 1974, C

743. San Joaquin Bank
(Bakersfield, CA.)
6 ¼", T8, S, c. 1970s, D
Name on sides of base. Clock on all sides.

744. Tower Federal Savings
(Western Spring, IL)
5 7/8", T9, S, c. 1970s, B

745A. Worcester Federal Savings
Worcester, (MA.)
4 5/8", T1, S, 1952, C
"Your Doorway to Thrift. 75 Years of Service - 1877-1952" Slot in top also. "Worcester - 22 Elm Street" "Springfield - 1243 Main Street"

745B. Worcester Federal Saving
on Marble Base (3/4") – C
Pen Holder. Same bank and wording as #745A

746. Worcester Federal Savings
Worcester/Springfield, (MA)
5 3/8", T4, S, 1957, C
"Our 80th Year - Founded in 1877"

747. Allegheny Mountain Tunnels
Gallitzin, PA.
3 1/8", T9, G, c. 1970s, C
Gallitzin S&L Assn. - Tunnel -
"1854 and 1904 – Elevation 2167 Feet"

748. Hoosac Mountain Tunnel
Adams, Mass.
3 ½", T9, G, c. 1970s, B
Adams Cooperative Bank -
Paper sticker on bottom

Chapter 12
Traps, Keys, Variations, and Boxes

Traps

The trap on a bank is where the coins are re-oved from the bank and, in many cases, also where ıe coin is inserted. The coin slot often is found in ıe trap, but may be located on the top, back, side, r a separate place in the bottom next to the trap. 3 different traps are identified in this chapter. Some aps are one-of-a-kind on a particular bank and ot used again by Banthrico. These are usually the ewer plastic banks. The steel and white metal traps ere used repeatedly by Banthrico in a clear evolu-on through the years.

The traps are numbered T1 through T43. The ank traps were numbered as they were discovered y the author. They were placed in chronological rder at first as much as possible. For example, the 25 is a very early trap found only on the autobanks ut was done later for the book and therefore has a igher number. The traps are all pictured and num-ered with a basic definition in the captions. The ank that goes with the trap is also given with its ook number. The bank that is listed with a particu-ır trap does not mean that trap was used only on ıat bank. Steel, white metal, tin and plastic traps re generic and not specific, as a general rule. There re a few exceptions (#781), but they are rare. anthrico used these 43 traps (there may be one or two that I missed) on millions of their banks. It is not unusual to see several different traps on the same bank design made through the years, especially if the bank was popular. The Abraham Lincoln Bust bank is a good example. It had at least four different traps – T1, T4, T8, and T9 – over time. You may have a different one in your collection. By the same token, one trap may have been used on a hundred different advertised banks and producing several hundred thousand Banthrico banks. An excellent example of that is the T4 rectangular steel hinged trap that was used more than any other trap Banthrico produced. The T4 trap appears on people, animal, trophy, mascot, transportation and architectural banks, to name just some.

All traps are located in the bottoms of the banks, unless indicated otherwise in the trap description. The slot is also located in the bottom trap, unless described and shown in another place. Most square base plates are held in place with four screws, rivets, brads or solder. The round base plates are usually held in place with three screws or rivets.

The many variations that occur in each trap, such as the location and slant of the coin slot, bill hole; type of screws, brads or solder; no slot in trap; type of saw teeth; no teeth; spring wire guard; color; size; metal plate configuration; etc. are not given "T" numbers. That could produce several hundred different trap types. The pictures and text cover some of these trap variations. Banthrico was quite prolific in trap types and variations.

The approximate dates the traps were first made is not easy to determine. Some traps started were continued for several years following their introduction. The following chart gives the approximate dates for each trap. Several of the traps were probably made concurrently. Trap numbers T34, T43, T20, T23, T25, T6, T7, T1, T2 and T3 are definitely some of the oldest traps made by Banthrico. Traps T20 and T23 do not have company names on them, but are believed to have been made by Banthrico. Traps T20 and T25 (removable key locks) appear to be forerunners of the removable key lock T1 and T2 traps. Traps T23 and T43 (hinged) appear to be harbingers of the hinged T4 and T5 traps. Traps T20 and T23 are on the older and smaller bust banks, like F.D. Roosevelt (T20), George Washington, and Abraham Lincoln (T23) that have Banthrico characteristics. Many of the T6 traps have no company name on them, but other research reveals they are Banthrico. The majority of the T1, T2, T3, T4, T5, T7, T8, T9, and T25 traps do have the Banthrico (or their predecessor) names on them.

Banthrico Traps: Approximate Dates First Made
(Some Traps are Companies They Purchased)

1920s	1930s-40s	1950s-60s	1970s-80s
T34	T1	T4	T9
	T2	T5	T10
	T3	T8	T11
	T6	T13	T12
	T7	T15	T14
	T17	T16	T19
	T18	T22	T33
	T20	T24	T36
	T21	T26	T37
	T23	T27	T38
	T25	T28	T39
	T30	T29	T40
	T41	T31	
	T42	T32	
	T43	T35	

Banthrico was constantly evaluating their bank traps and looking at ways to improve them while making them more economical. A definite evolution of the bank traps occurred, especially starting with T4. It was a stamped steel, hinged trap made by Banthrico in a rectangular shape on a steel plate. They also made a hinged trap (T5) in white metal on a steel plate. Next you see Banthrico making a rectangular white metal trap in a steel plate (but not hinged) (T8). After T8 came T9, the same steel plate, but with a tin rectangular insert trap. Following T9 came T10 with a plastic rectangular insert on a steel plate. If you will review the steel plates for T4, T5, T8, T9, and T10 you will see they are all the same with only a minor change on the hinge area found on T4 and T5. They all have the rectangular opening in the steel plate. Banthrico had their own stamping machines so they stamped out the steel plates, traps, and keys that opened them. They also cast the white metal traps (T5, T8, T13 are examples) and keys. The white metal traps and keys were zinc and aluminum (ZAMAC) similar to the bank composition.

If you are trying to date a bank, be sure to examine the trap. The trap is a definite indicator of the circa age of the bank (unless the trap has been changed from one bank to another which occurs). An Abraham Lincoln bust with a T1 trap is definitely older than an Abraham Lincoln bust with a T9 trap. An autobank with a T25 trap is definitely older than an autobank with a T4 trap or a recast autobank with a T9 or T10 trap. The trap is not the only indicator of a bank's age, but it is certainly one of the indicators.

Many of the traps had no coin slot in the base trap and the slot is found on the top, back or side of the bank. The captions alert the reader to this fact. Sometimes the coin slot is located in two places (on the bottom and on the top of the bank) and that will be indicated also. You will also note that many of the traps have a small hole in them (not a bill hole) for putting a string through the hole and attaching the trap key to the bank so it won't be lost.

Many Banthricos have literally "lost their bottoms" due to lost keys and base plates removed for coins and not replaced or removing one base plate and/or trap for another bank that needs one (since the traps/base plates are interchangeable if they are the same size). This is why you often see a Banthrico with a converted trap that is a steel, tin, aluminum, or white metal plate with a coin slot (similar to T6). If a bank is found without a trap or base plate, it will probably receive a converted trap (plate and slot).

Many early steel and plastic house bank traps have "BT Corp" or "Banker's Thrift Corp." on them, but the company was owned by Banthrico (starting in 1931) and they elected not to change the name at first. Later, the same bank trap was made with "Banthrico Corp." on them. When Banthrico moved to snap-in traps (T9 & 10), customers were given a choice of lock or snap-in traps for their bank purchases. The lock traps were more expensive because they required the stamping or casting of a lock mechanism and a key.

The famous engraved name plate "Copyright 1974 Banthrico Chicago" was placed on the steel bottom plates starting in 1974 and continued until the company closed in 1985. There are exceptions to the "Banthrico 1974" name plate that have no engraving; the First National Bank of Chicago is one of them. T8 traps (1950s-60s) can be found with the "1974" steel plate since Banthrico put T8 traps already stocked with the new "1974" steel plates. The Old Capitol Bank & Trust Co. building is an example. When Banthrico was sold to Toystalgia/Valley Casting, the steel base plates were included in the sale. Valley Casting wisely used these "1974" steel plates with their banks until quite recently. They probably depleted their stock. Therefore, people may think they have a "Banthrico" bank that actually is a Valley Casting bank. The "1974" base plate is usually found with the T9 and T10 traps, but may also be found with the T8 trap. Banthrico's rectangular plastic plug insert (T10) is dark dull gray while the Valley Casting plastic plug is shiny pewter gray. If you find a white metal bank that has "Banthrico 1974" on the base plate with Phillips screws in the base plate, it is a Valley Casting. Banthrico did not use Phillips screws. Valley Casting did use other screws in their banks at first, probably from the Banthrico buyout.

Keys

The keys that are shown open a number of the bank traps listed, but some of the bank traps are unaccounted for in matching a key. In picture #801 (L to R) the #3 steel key for T4 is seen the most because the T4 trap was the most used Banthrico trap. The #4 white metal key is the same as the #1 key in picture #805 (L to R). This key fits some five bank traps as listed and maybe others. The #5 white metal key (#801) is the same as the #2 key in picture #805. As you can see by the captions, the key fits T5 and T8. The white metal traps and keys were cast together in one mold by Banthrico. The #805 picture was enlarged so you can read the "Banthrico Chicago" on the #1 key and the "BII" on the #2 key. "BII" stand for "Banthrico International Industries." The other three keys in picture #801 fit Banthrico bank traps as indicated, but I'm not sure they were made for Banthrico traps. The #1 key is a "Crest" key and the #2 is a "Master Lock Co." key and the #6 key is unidentified. Banthrico could have used other companies in making keys for their traps, but I lean toward the theory that they made all of their own keys. Pictures #803 and #804 are also Banthrico steel stamped keys. The #803 picture is an early steel trap and key for the small plastic house. The #804 picture is an early steel key for the autobank series. There may have been other autobank keys. All the autobank traps (T6, T25 and T4) were steel traps. The T6 is a keyless trap. Some of the later autobank recasts had cast white metal traps (T8). Picture #802 showing three keys has only one key (#2-L to R) which is a Banthrico stamped steel key (for T4). The other two keys are made by other companies and may have been made for Banthrico traps or been ground down to fit the trap. Keys are difficult to find with the banks. As stated earlier, the rating of banks is not impacted by missing keys. Keys seem to wander away from the banks and never make it home again. Banthrico moved toward keyless traps in the 1970s and 1980s to eliminate the problem of lost keys and for economic reasons. Banthrico had stamping machines for the steel plates, traps, and keys and they also die-cast traps and keys in white metal, so this leaves little room for Banthrico cor tracting out keys for their banks.

Variations

There are a few variations listed here that hel explain more about Banthrico banks. Picture #80 shows the white metal banks as they looked whe first coming out of the mold and before they hav been electroplated, finished, and traps installed. Th other variations show the back of some banks tha have metal plates and raised metal in the bank; pape stickers from Chicago and British Hong Kong; an the reverse side of an interesting trophy bank (Hol Redeemer Church with key). The two date-registe ing container banks shown side by side give you top view of the banks, and the different size coi slots for dimes, quarters, and half dollars. The las variation (#812) is a building bank that did nc receive tender loving care. This is what happens t a white metal bank when it is not kept clean and i a dry location. This subject was discussed in dept in the Chapter 1, regarding Care and Cleaning.

Boxes

The 69 bank boxes are pictured #814 throug #883 and are arranged in the same order as th chapters. A few examples are given for each chap ter. Bank boxes (and other toy boxes) had a hig mortality rate prior to 1950 because people di not realize their value. Banthrico (and their prede cessors) used cardboard boxes, the earlier boxe were heavier than the later boxes. The early boxe were usually plain and the more colorful boxes ar later. The "SAV-O Clown" box is one of the mos colorful that Banthrico marketed. The Eisenhowe box states that the bank is a "statuette bank" an displays the same key that was shown in pictur #803. The red, white and blue patriotic box c General MacArthur is quite attractive. The masco bank boxes are quite elusive. The "Jack-in-the-Bank (not box) is another colorful box as is Donald Duc and Mickey Mouse. The Jack box is more than 2 years older than the Donald and Mickey boxes. Th Santa box picture shows the rare sack of small toys Note on the Santa box the picture of Franklin wit the inscription "Thrift Savings since 1914. Banthrico was using the beginning date of th Strongheart Co. as their beginning date. The Dachs hund box is one of the best made and creative boxe that Banthrico did.

Some boxes have "Banthrico bank names" o them, but a different name has been used in th book that was more appropriate. The "Toyland Bank box with handle and flocked bank is included i Chapter 13. The kepi caps of the North and Sout Armies came in gray and blue as indicated on th box. The word "kepi" means a military cap with round flat top sloping toward the front with a visc (or bill). The Date Changer bank is still in the bo with a window to see the bank (#848). The #85 box shows six classic autos produced by Banthrico Photos #851, #856, #857, and #858 also show "Antique Autos," but the 1977 Volkswagen pic tured has a hard time fitting into that category. Th 1908 Cadillac Limousine (#852) is a good exampl of the heavier cardboard boxes used and #853 an #854 are examples of the lighter, thinner carc board boxes. The Futuramic box was also made ou of heavy cardboard and the bank is extremely diff cult to find with the box. The maroon sticker o the Mormon Wasp box announces the formation c he "Antiquo" Division within Banthrico at the :oosevelt Road location in the late 1970s and early 980s for gift shop customers.

The "American Home Bank" box (#869) is the xact shape of the bank and done in color depict-ng a home which makes it most attractive. The Drive-In Coin Bank box provides the instructions for working the mechanical bank. The International Building box is quite historical and attractive, giving all the details of the dedication ceremonies. The Union Savings steel bank and box and the Waterbury National Bank and box are fine examples of heavy boxes with two sections - the box and the cover for it. These boxes are the plain "Cadillac boxes" of Banthrico. The Statue of Liberty box (three sides) is loaded with information for the collector.

'49. T1-Steel Plate with Round
teel Key Lock
eated Indian #28

750. T1-First Fed S&L - Detroit
Left-Lead Bank (Soldered) #592A
Right-WM Bank (Brads) #592B

'51. T1-Variation – Key
/liami Beach Fed S&L
:arly Version #656

752. T1-Variations – L to R (#1&#2)
Eisenhower #16-Fat Pelican #154-
T2-Steel Pl with Rd Alum Key Lock (#3)
Cont. Assur Co. #565

753. T3-Steel Plate with Key Lock
L to R-Furniture Cab. #177
N. P. Barrel #262
Pig with Big Ears #164

754. T3-Variation
Carved Coconut #200
With Steel Key

755. T4-Steel Plate with Rectangular
Steel Hinged Key Lock.
L to R – LaSalle #32 - Idaho Pine Log #246

756. T4-Variation
Saw Teeth Guard
Idaho Potato #206

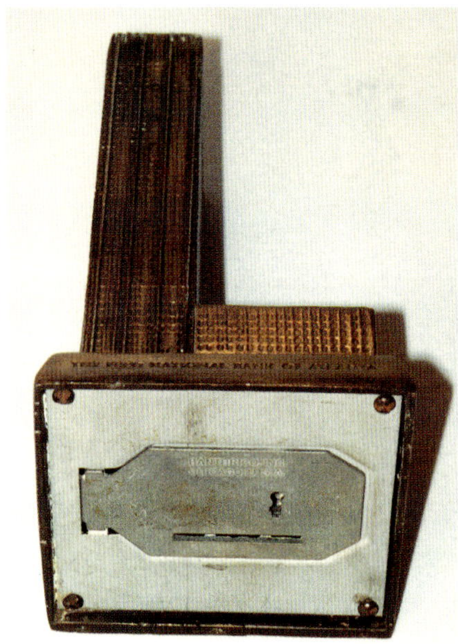

757. T5-Steel Plate with Rectangular
White Metal Hinged Key Lock. First
N. B. of Atlanta #603

758. T6-Steel or Tin Plate
L to R – Iowa Hawks #77
Astronaut on Moon #4

759. T6-Variation
Instructions on Base
Ben Franklin Bust #20

:0. T7-Steel Plate with
ectangular Steel Key Lock.
ouse with Dormers #635

761. T8-Steel Plate with Rectangular
White Metal Key Lock.
Goldome Building #627

62. T9-Steel Plate with
ectangular Tin Insert.
eated Schnauzer #140

763. T10-Steel Plate with
Rectangular Plastic Insert
Tractor #503

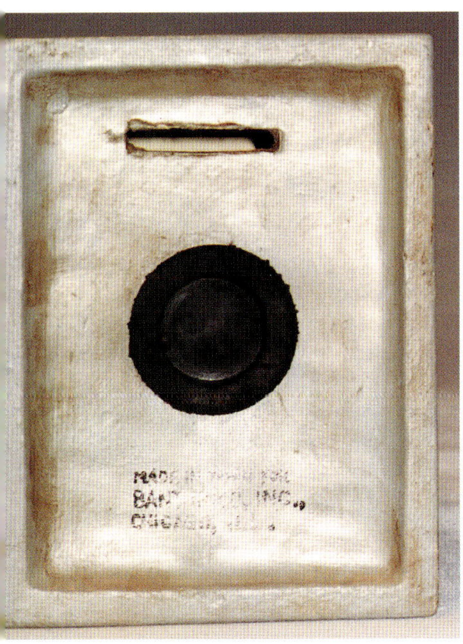

764. T11-Slot and Rubber or
Plastic Stopper in Bottom
Arsenal – Composition #533

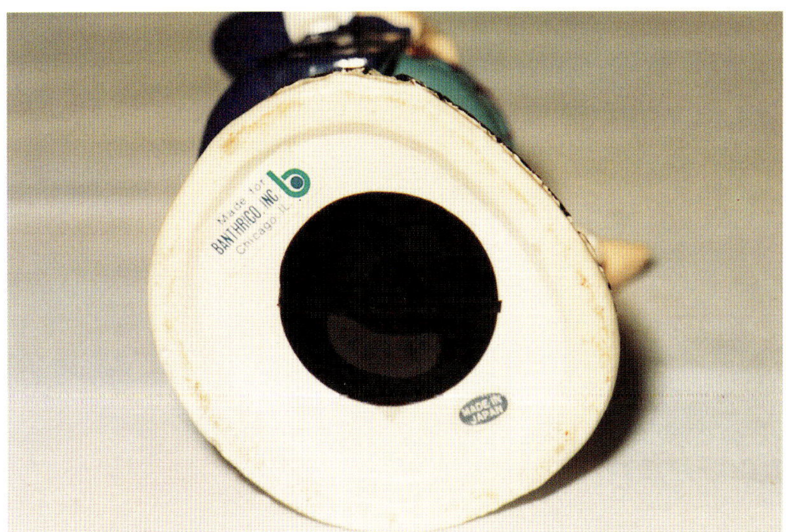

765. T12-Slot in Top-Rubber
or Plastic Stopper in Bottom
Ben Franklin #21

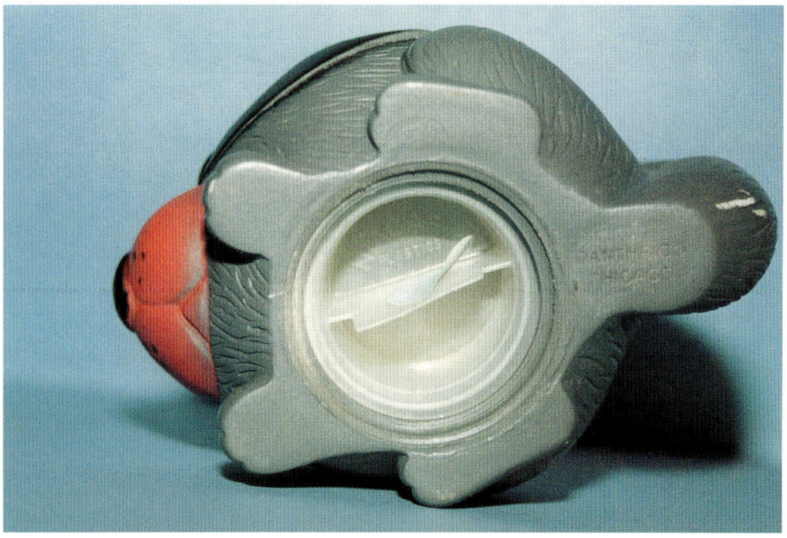

766. T12-Variation
Plastic Stopper
Raccoon #169
"Made for Banthrico Inc. - Il.
Made in Japan"

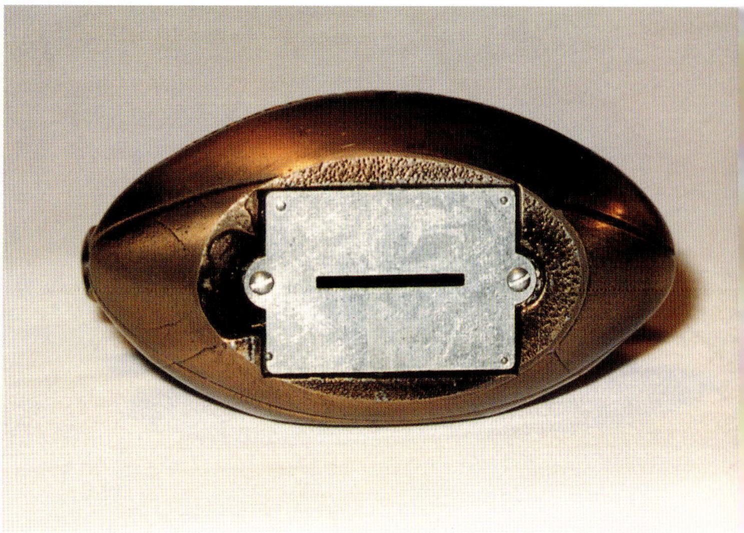

767. T13-White Metal Plate
with Screws.
Arkansas Razorback #64

768. T14-Wood Base with
Screws – Slot in Back.
Robert Kennedy Bust #30

769. T15-Key Lock in Back
Opens Bank Into Two Pieces.
Plastic Mechanical Clown #10

770. T16-No Trap – Solid
Chalk Bottom – Break to Open.
Sleeping Santa #121

771. T17-Round Steel Key
Lock in Top with Slot.
Nickel Plated Barrel #263

72. T18-Side Key Lock
emoves Steel End.
alculating Cylinder #292

773. T19-Round Plastic Twist
Out Trap.
Mickey the Magician #116

74. T20-Nickel Plated Steel Plate
nd N. P. Round Steel Key Lock.
D. Roosevelt #47

775. T21-Round Steel Plate with
Round Steel Key Lock-Slot in Top.
Wooden Oak Barrel #174

76. T22-Key Lock Removes
astic Top with Slot.
mbrella #194

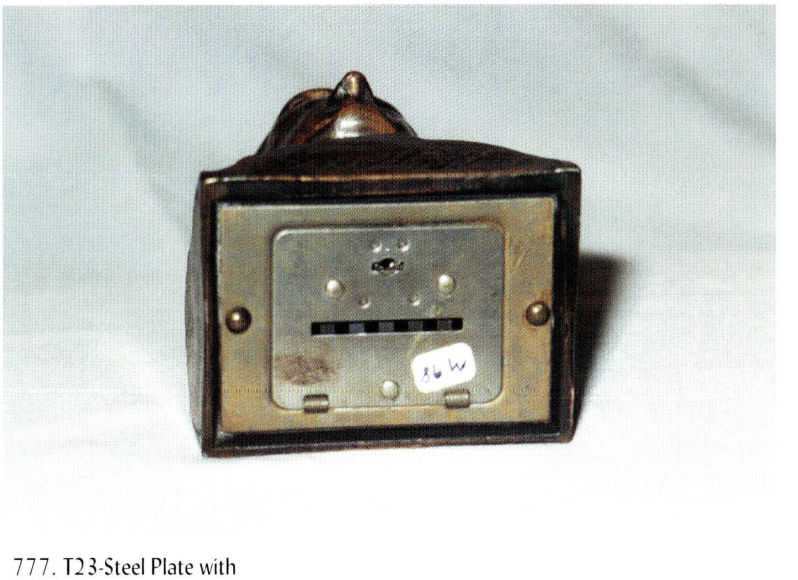

777. T23-Steel Plate with
Rectangular N. P. Steel K. L. w/Hinges
George Washington Bust #58

778. T24-Round Plastic Plate with
Key Lock and Slot in Top
Large Metal Barrel #261

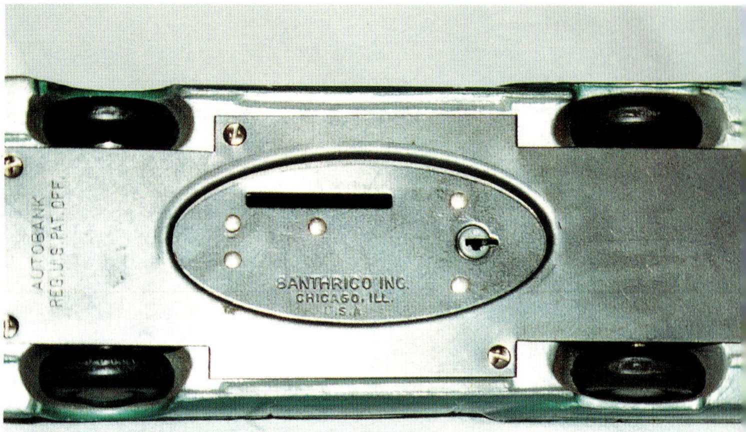

779. T25-Steel Plate with
Oval Steel Key Lock.
Packard Clipper #401B

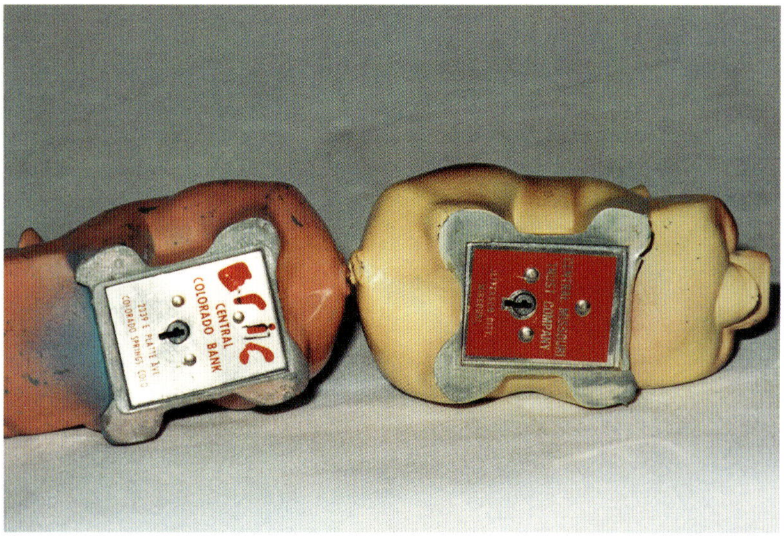

780. T26-Aluminum Plate
With Key Lock. (L to R)
Happy Colorful Pigs #158, #157

781. T27-Spicket (Key) Opens
Barrel End
Whiskey Barrel #264

782. T28-Plastic Shirt Slides
Down in Front
Top Hatted Pig #166

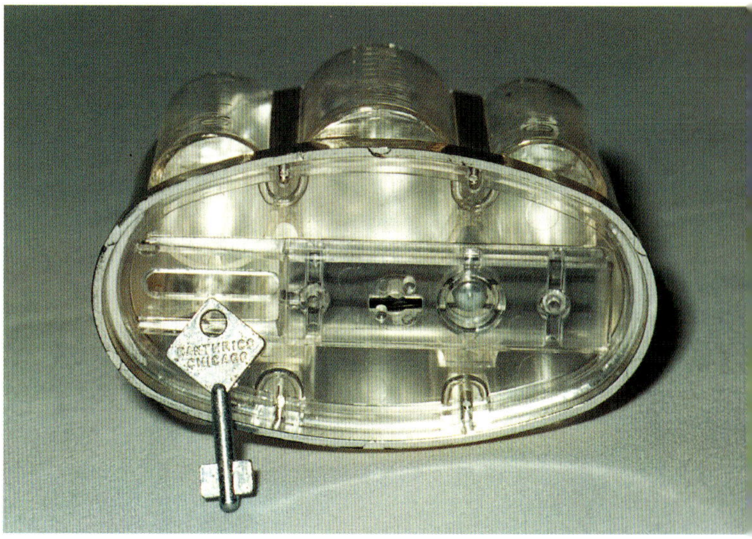

783. T29-Plastic Key Lock
In Bottom.
Sumshine Bank #285

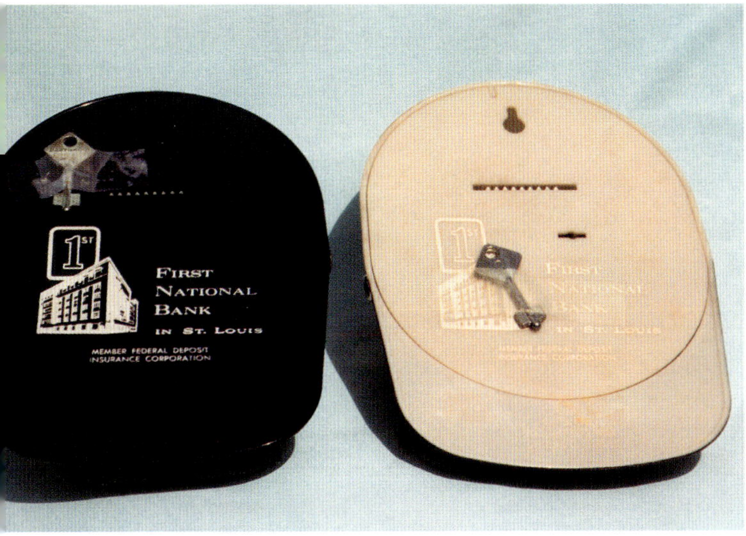

84. T29-Variation
to R – Union Army #254
outhern Army #253
ote "ZAMAC" Key for
pening bank

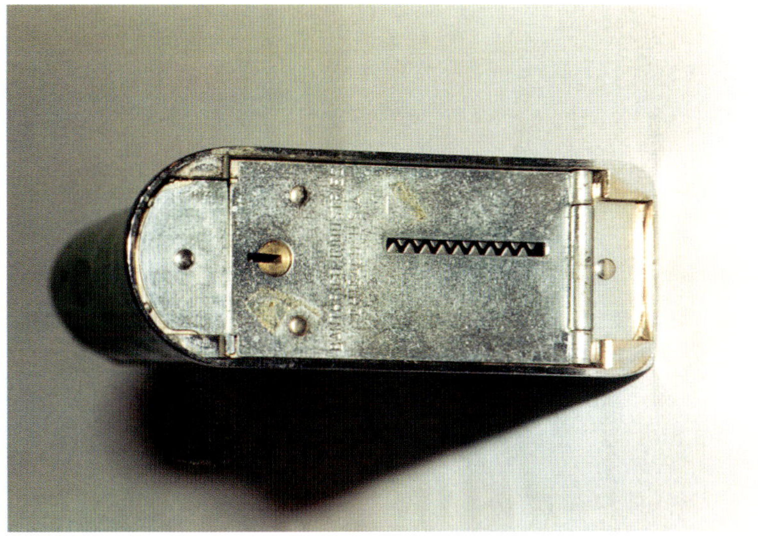

785. T30-Steel Plate with
Hinged Key Lock.
Chromeliner #269

86. T31-No Trap – Two
des Pried Apart to Open.
orthwestern Tower – Plastic #739

787. T32-Plastic Back Plate
with Key Lock and Inward Slot.
Date Container #318

88. T33-White Metal or Plastic
lug in WM Bottom – Slot in Top.
teamboat on Water #958

789. T34-Slot and K. L. in Side
Of Steel/Tin Disk.
Jefferson #328 – Reserve Bldg. #339

790. T34-Variations
Key Locks and Slots
Merchants #331-Park Savings #332

791. T34-Variations
Deep Pocket Bank
Northern Bank &
Trust #353
Coin Slot

792. T34-Variation
Key and Currency Holes.
Northern B & T #353

793. T35-Hat Removed for
Coins – Slot in Hat.
Hopalong Cassidy #26

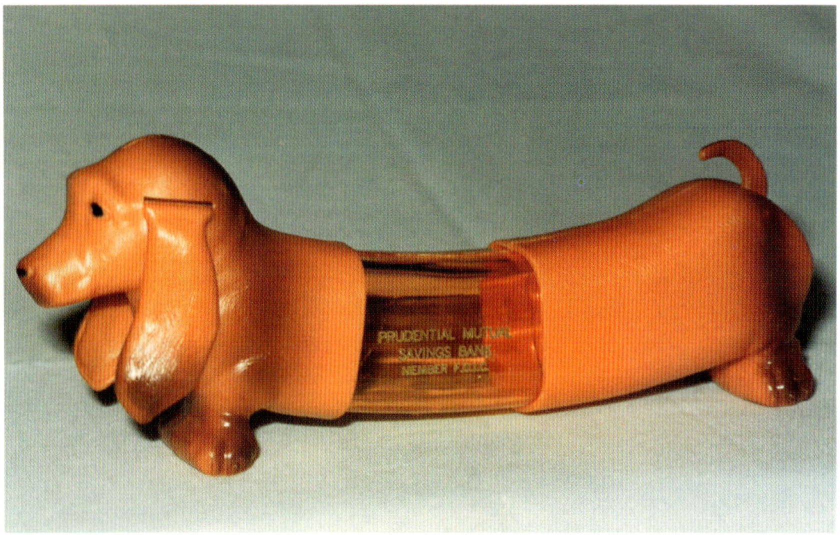

794. T36-Pull Two Sections
Apart at Center for Coins
Dachshund Dog #136

795. T37-Tin Lid with
Slot Removed for Coins.
Tin Container #274

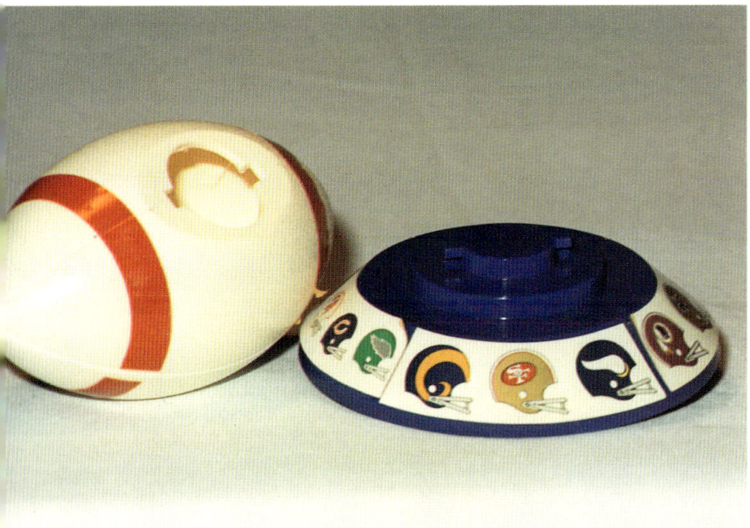

96. T38-Plastic Twist Off
ase With Slot in Top
FL Football #93

797. T39-"Squeeze Ends at
Arrows to Open or Close".
Small Suitcase #187

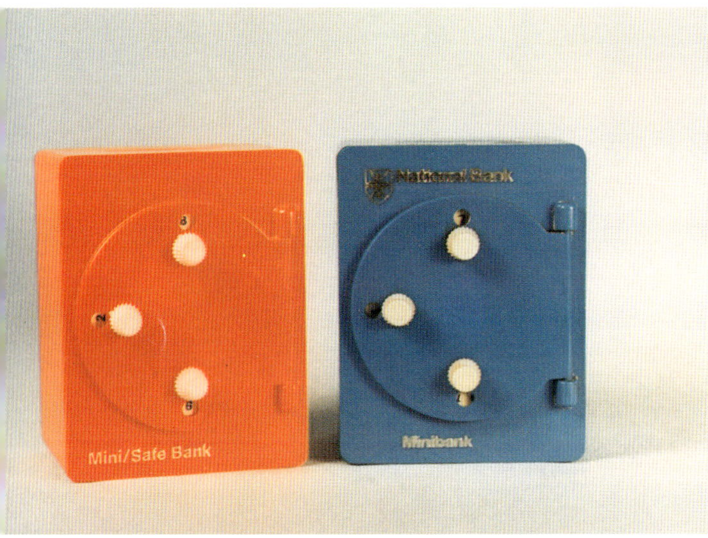

98. T40-Combination Door
pens Bank. (L to R) - Mini/Safe
Minibank Safe #282, #281

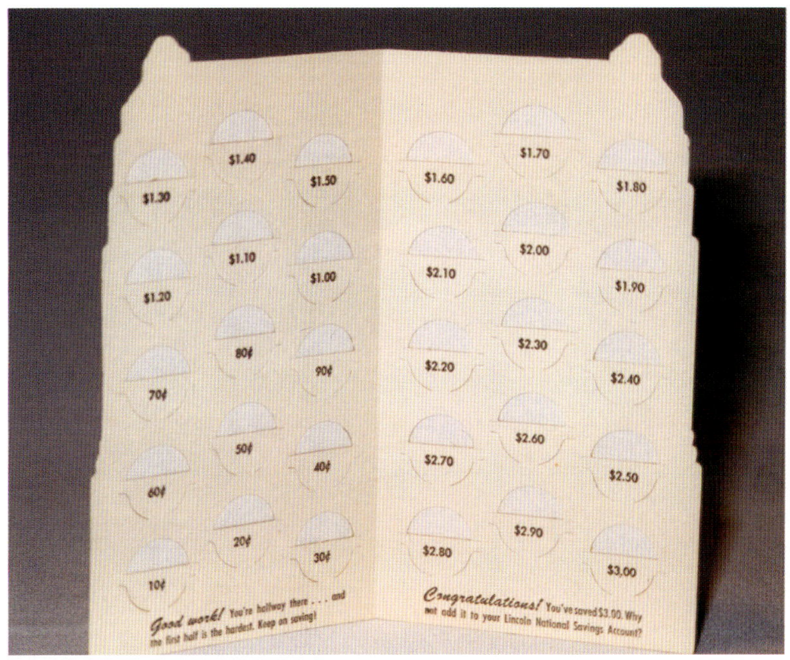

799. T41-Coins Placed in
Pockets Inside Folder.
Lincoln National Bank #351
Note cumulative amounts up to $3

00. T42-Remove Child
om Chamberpot for Coins
hild on Chamberpot #884

800A. T43-Square Steel
Key Lock Hinged Plate
MacArthur Bust #924

801. Trap Keys (L to R)
#1-T3 - #2-T3 - #3-T4
#4-T15 - #5-T5, T8 - #6-T25

802. Trap Keys (L to R)
#1-T25 - #2-T4
#3-T1, T2, T4

803. Trap Key
Steel Key – T3
Industrial Bank #638

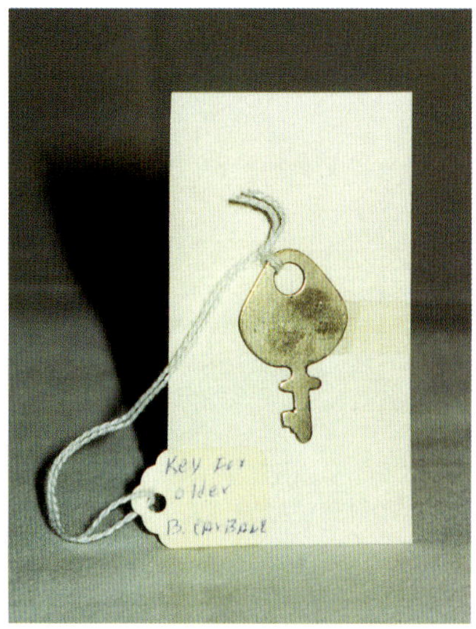

804. Trap Key
Steel – T25
Autobanks

805. Trap Keys (L to R)
#1-T15, T22, T24, T29, T32
#2 – T5, T8

806. White Metal Banks
Before Electroplating
and Finishing

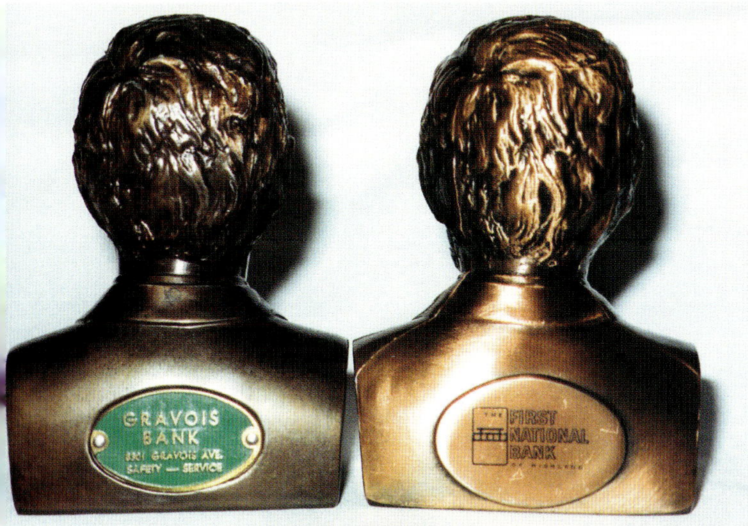

808. Franklin Variation
Back Side
Metal Name Plate

07. Lincoln Variations
ack Sides – Name Plates
letal Plate – Raised Metal

09. Holy Redeemer Church
ack Side with Key (T4)
lot in Top #218

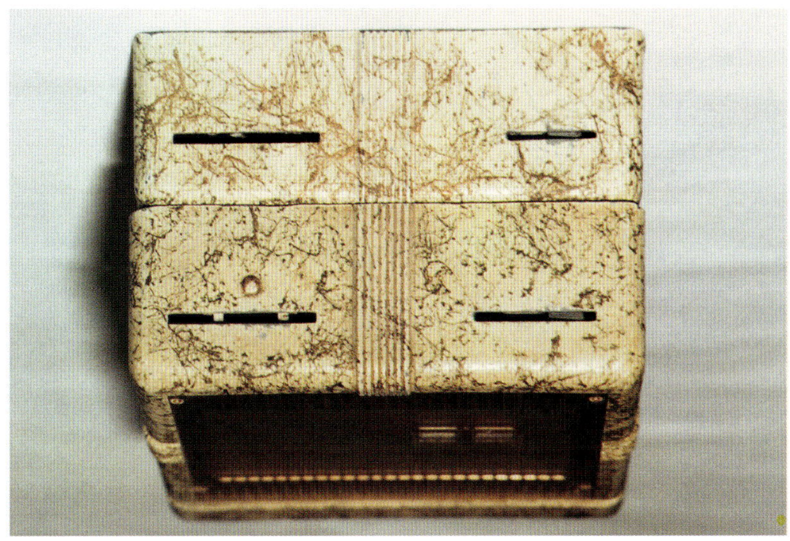

810. Date Registering
Containers (2) – Top View –
Coin Slots #316, #317
Top: "Press here to close"

811. Banthrico Paper Sticker
Found on Early
Banks – Paperweights

812. Citizens National Bank
Heavily Oxidized #556
"Fair" Condition Example

813. Base of Bozo the Clown
"Made in British Hong Kong
for Banthrico, Inc. "

814. Clown Box
Sav-O-Clown Bank
Old C. I. Bank Pictured

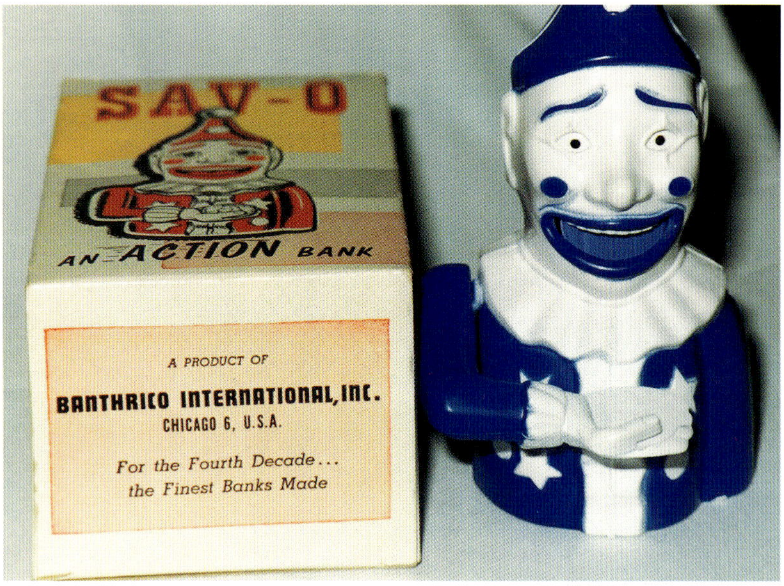

815. Clown with Box
"For the Fourth Decade
the Finest Banks Made"

816. Sav-O-Clown Box
Operating Instructions

817. General Eisenhower with Box
"Statuette Bank"
Key for T3 Trap

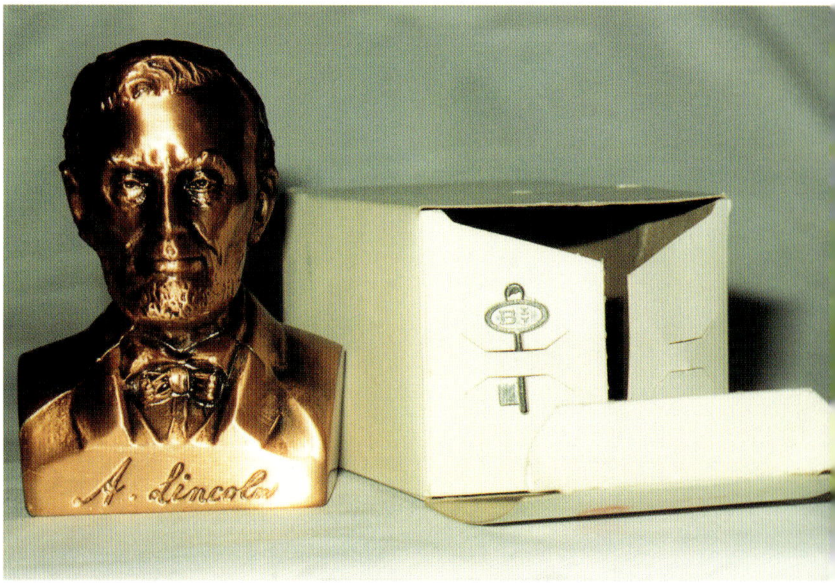

818. A. Lincoln with Box
Key(s) in Box Lid

19. Bust Bank Box
ranklin – Indian – Kennedy
ncoln – Washington – Other

820. MacArthur Banks with Box
Two Variations
"Save for Your Country's Future"
"The General Bank"

21. Missouri Tiger with Box
lso Bear Box

822. "Jack-in-the-Bank" with Box
Plastic Mechanical
Takes All Sizes of Coins
Put coin in slot and out comes Jack

23. "Jack-in-the-Bank" Box
ront View
ank Talks
Thank You" sound. Stretches
ut arms

824. Donald Duck Box
Plastic Mechanical
Takes All Sizes of Coins

825. Mickey Mouse Box
Walt Disney Character
"Copyright 1981"

826. Santa with Box
Back Side
Note Franklin Head/Box
"Thrift Devices Since 1914"

827. Santa on Roof
Sack of Small Toys
Worcester – Springfield (MA)
Worcester Fed S&L Assn.

828. Cat Head with Box
Glass Eyes
Gold Plated

829. Dachshund with Box
Pull Apart to Enlarge
Ears Move When Coin Inserted

:30. Dachshund Box
'ats on Both Sides/Roof
Banthrico International Inc. "

831. Red Pig with Box
"Blubber Boy"

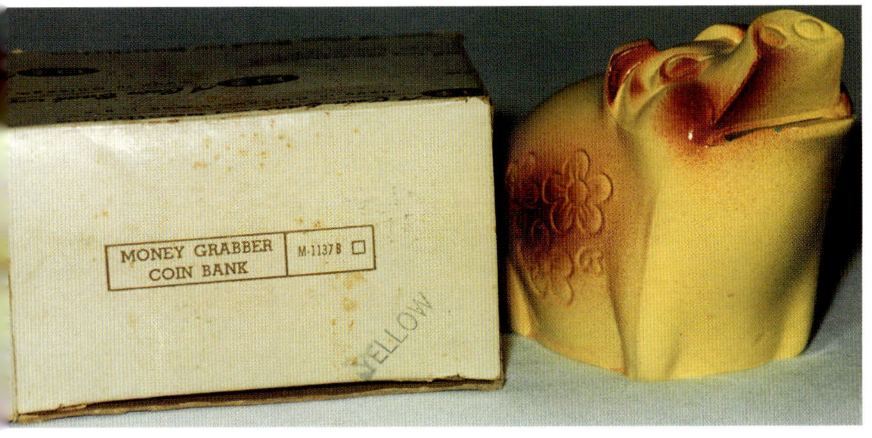

:32. Pig with Mouth Open
Money Grabber" Box

833. Poodle with Box
"b – the Coin Bank People"

834. Raccoon with Box
Flocked Vinyl Bank
"Toyland Bank"

835. Jewelry Box with Box
Came in Black – Brown

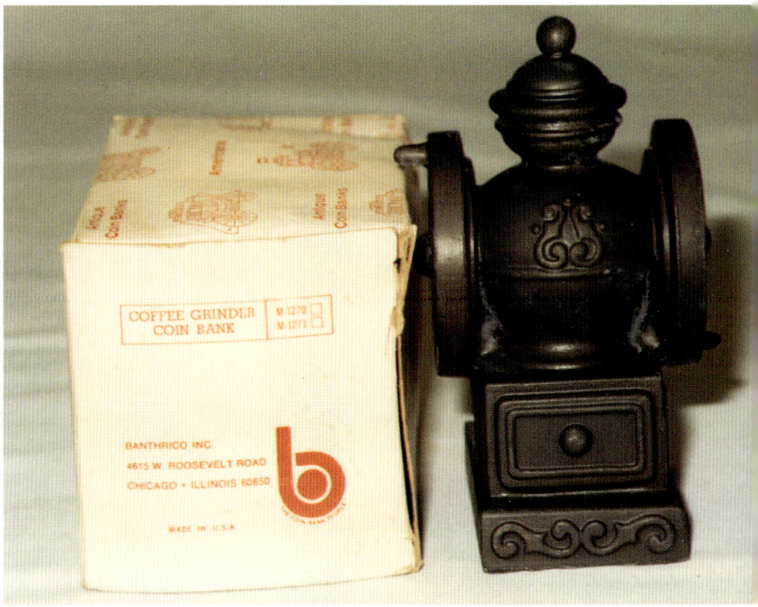

836. Coffee Grinder with Box
Roosevelt Road Location

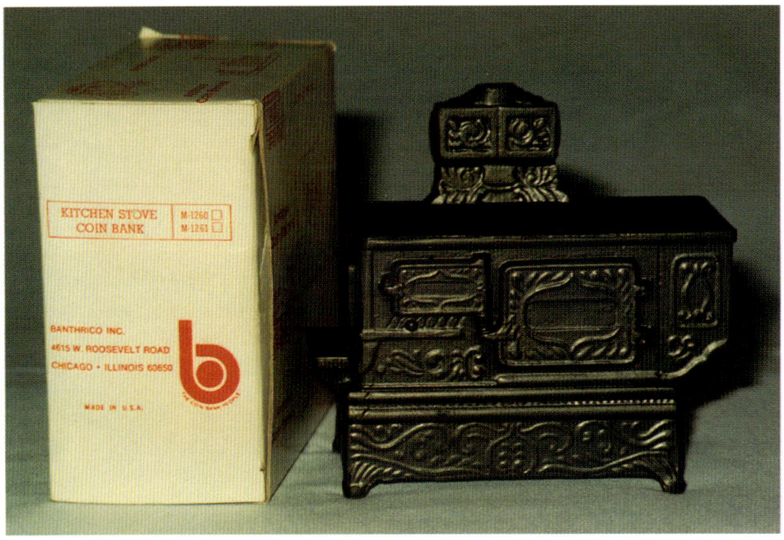

837. Cook Stove with Box
"Kitchen Stove"

838. "Americana" Series Box
Cook Stove
Series of 1970s Banks

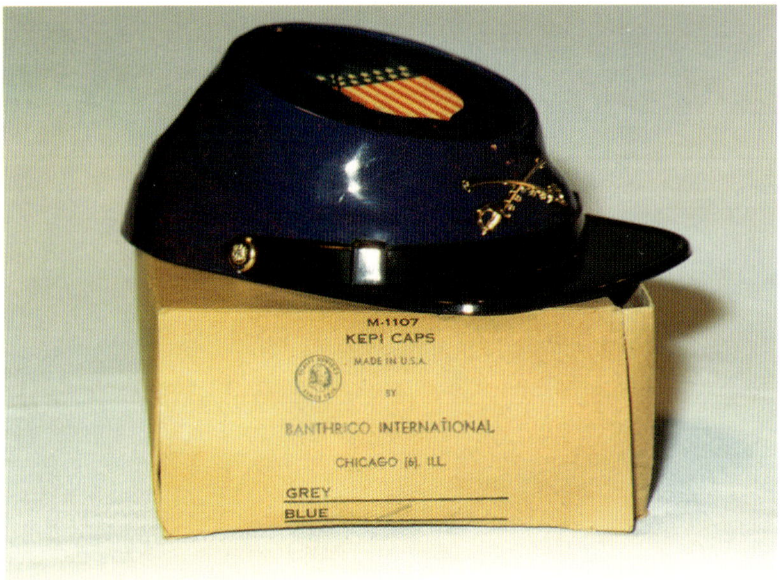

839. Union Army Calvary
Cap with Box
"Kepi Caps"

840. Medium Metal Barrel
with Box
"Stave Barrel Bank"
First Tr & Svg Bk - Cedar Rapids, Ia
"The Friendly One"

841. Large Metal Barrel
with Box – T24 Key
"Koin Keg Bank"

842. Nickel Plated Barrel
with Box
"Staveline Barrel"

843. Oval Container with Box
Nickel Plated

844. Flash Cube with Box
"Photo-Magic Bank"

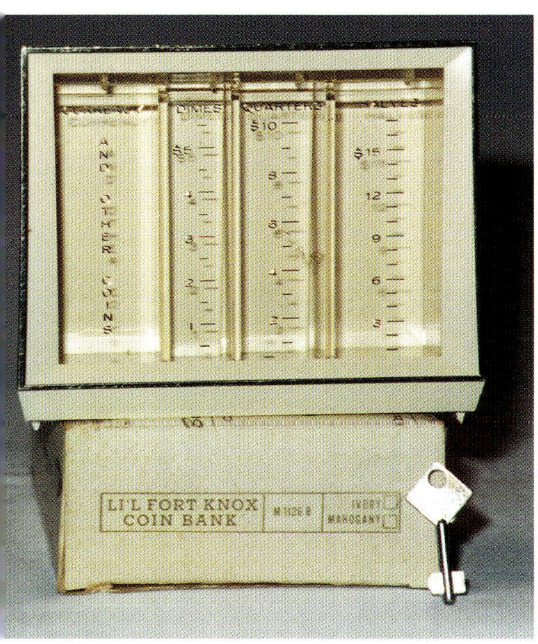

845. Li'l Fort Knox
with Box and Key
"Pennies-Dimes-Quarters-Halves"

846. Sumshine Bank
with Box
Plastic Bank and Trap
Plastic coin compartments for
different size coins.

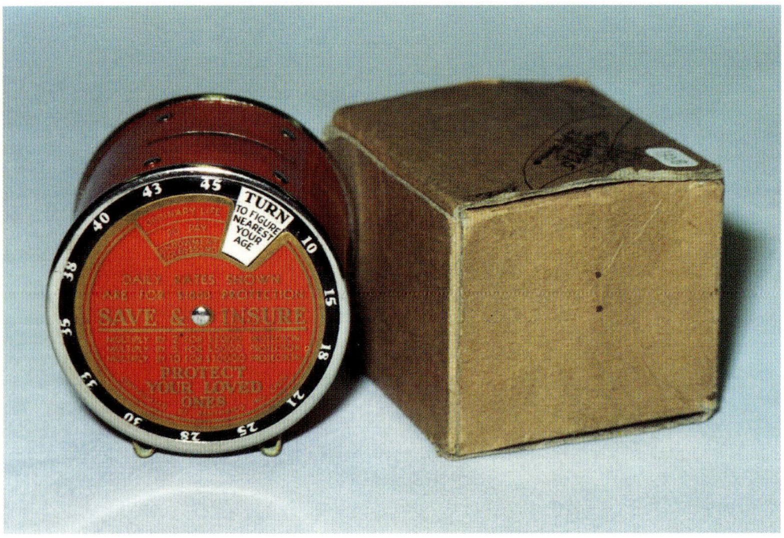

847. Calculating Cylinder
with Box
Adjustable Front Dial

848. Date Registering Container
in Box
"Copyright 1955"
"As the days go — your savings grow!"

849. Date Registering Container
with Box
"Daily Dater" Bank

850. Banthrico "Classic
Autos" Box
Six Autos Displayed

851. "Classic Autos" Boxes
Two Side Views

52. Cadillac with Box
908 – Non Moveable Wheels

853. Chevrolet with Box
1950 Coupe
Recast from Autobank

854. Corvette with Boxes
"Classic Car Series"
"Quality Products Since 1914"

855. Chrysler with Box
1946 Town & Country

856. Banthrico's Antique
Autos Box
Five Autos Pictured

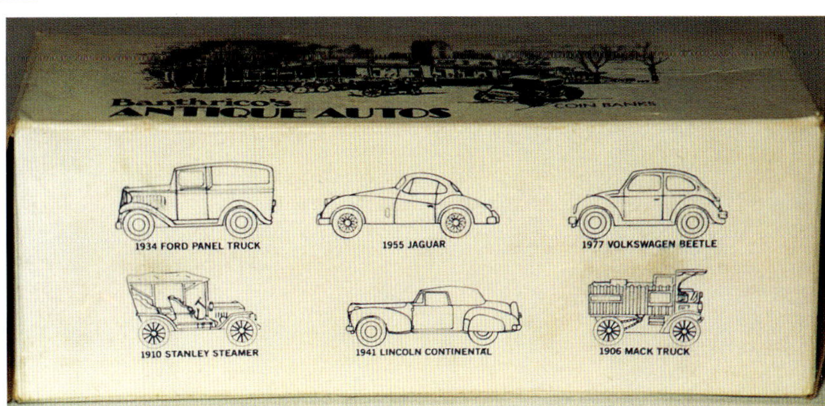

857. "Antique Autos" Box
Six Autos Displayed

858. "Antique Autos" Box
Six Autos Displayed

859. Ford Sedan with Box
1926 – Non Moveable Wheels

860. Model A Ford Roadster
with Boxes-1929 – Plastic
Bag Cover Under Car

861. Futuramic Autobank
with Box
"Copyright 1956 – Banthrico"
"An all metal, scale model coin bank"

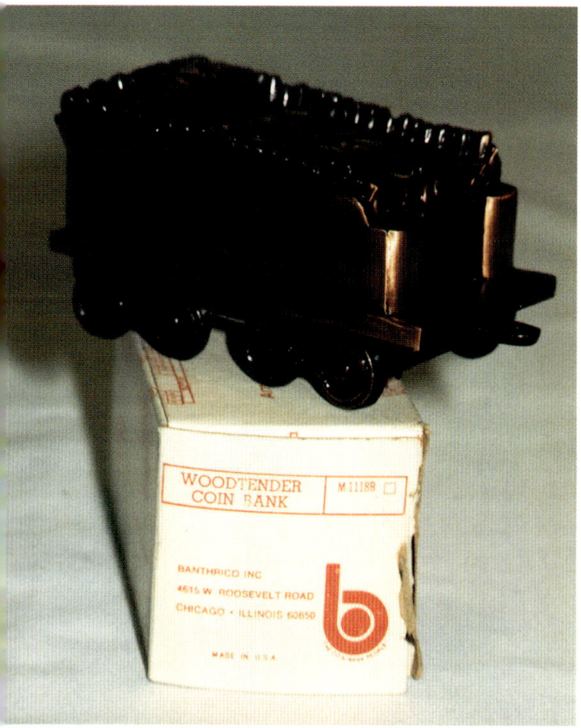

62. "General" Wood Tender
ith Box

863. Lincoln Continental
with Boxes - 1941
"Classic Antique Autos"

64. Marmon Wasp with
ox – 1911
Antiquo" Gift Division

865. Pierce Arrow
with Box – 1917
No ID on Bank

866. Popcorn Wagon
with Box
"Made in U. S. A. "

867. Stanley Steamer
with Box – 1910
Soft Top

868. Jeep – U. S. Mail
with Box – Bldg. Like
Mountain Springs Hotel

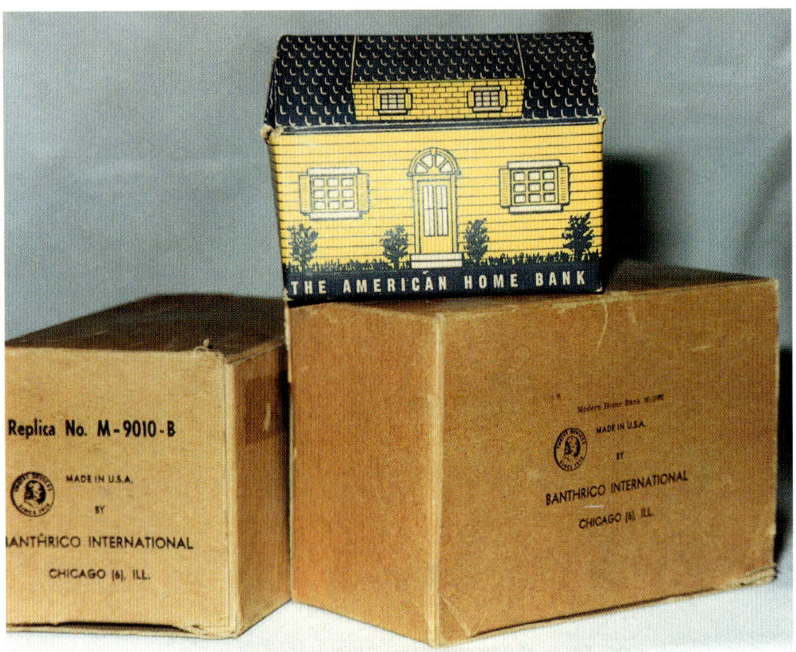

869. "American Home Bank"
St. Paul Fed. S&L (M-9010-B)
"Modern Home Bank"

870. Box Bottom
"American Home Bank"

871. "American Home Bank"
Earlier Box

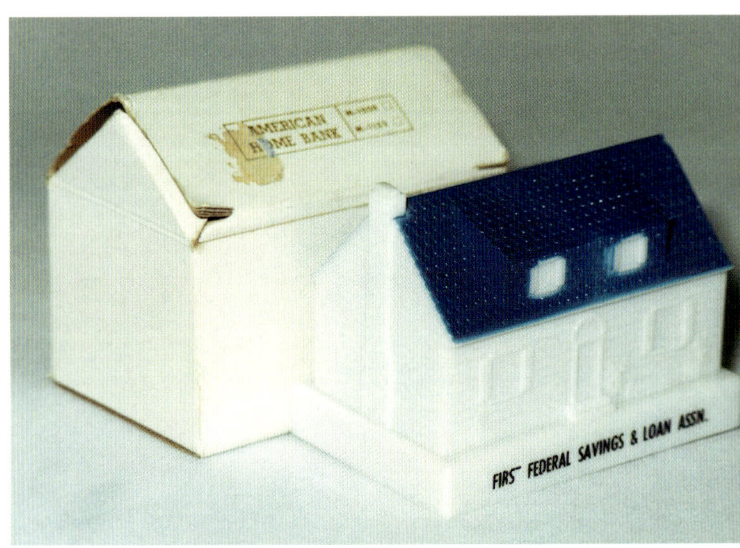

872. "American Home Bank"
Later Box

73. Brookfield Federal Savings
ith Box

874. Drive-In Bank
Box
"Heavy Duty Plastic"

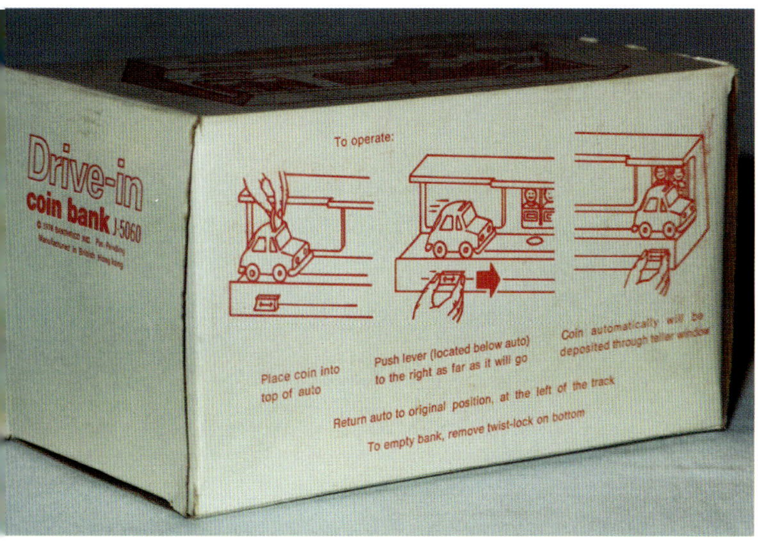

75. Operating Instructions
n Box
rive-In Bank

876. East Alton S&L
with Box – Key

877. First National Bank
of Chicago with Box

878. International Building
with Presentation Box
"Official Dedication Ceremonies"
"May 31, 1980 - World Council of
Credit Unions - Credit Union Nat.
Assn."

879. Union Savings
with Box
Two Piece Box

880. Waterbury National
with Colorful Box
Two Piece Box

881. Statue of Liberty
with Box – 8 ½"
Centennial Celebration

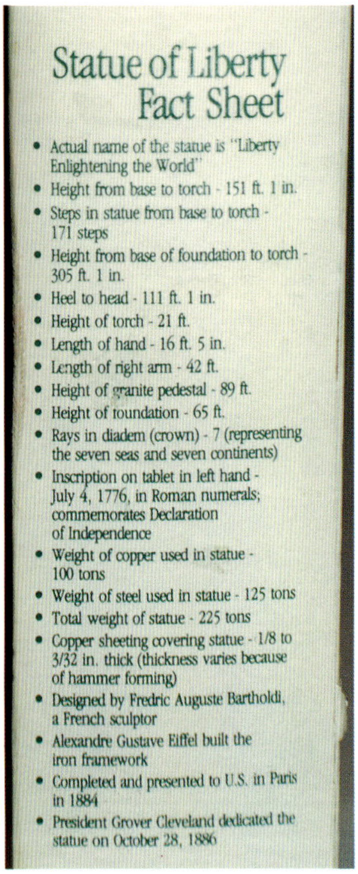

882. Statue of Liberty
Fact Sheet Box
Given to U. S. by France – 1884

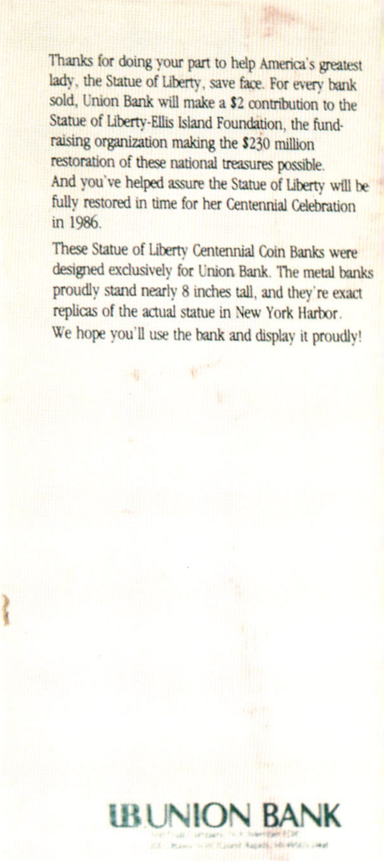

883. Statue of Liberty Box
230 Million Restoration Project
Designed for Union Bank & Trust Co.
Grand Rapids, Mi.

Chapter 13
Recent Discoveries and Information

The banks here were found after the chapters re written and numbers were assigned to the nks. They are "new finds" not in the earlier chaps. The bank pictures are arranged in the same der as the earlier chapters – people to architecral banks – and alphabetically. Some pictures ared late and are at the end of the chapter. Rently discovered information is also included.

There are several National Products, Inc. banks this chapter because Banthrico purchased their mpany in 1940 and continued many of their banks to the 1950s. There are no date markings on the ational Products banks to know when they were anufactured. Many of the banks have "National oducts" on their bottoms. The Child on hamberpot, the household items, cash register, pewriters, and boiler are all National Products nks. Other National Products banks are the Newpe Gasoline Pump, Sanitizor Vacuum Cleaner, tate Gas Range, Remington-Rand Typewriter, and yal Typewriter. They were advertised as "Miniares in Metal" (#1008, #1011, #1012, and 1013) and "your product or trade character hrunk" to scale, hand cast in indestructible aly." National Products made the minuteman as a perweight/desk ornament and they also made the nchburg National Bank and Trust Co. building nk which was continued by Banthrico. Banthrico ade a smaller minuteman and put it on a base as bank. National Products probably made other hite metal and/or lead building banks.

Some of the "Miniatures in Metal," other than nks, were produced by National Products and anthrico during the 1930s and early 1940s and llowing WWII for such well known firms as Interational Harvester, Diamond-T Motor Car Co., hite Motor Car Co., Reo and Studebaker trucks, uick, Chrysler, Studebaker, and Reo cars, estinghouse, Delco Motors, Metro-Goldwinayer, Hotpoint, Temple Radio, Wheeling Furnace orp., Burgess Battery Co. (Snuffer Muffler), gersoll, A.C. Spark Plug ("Cleenie," the laughing orse, in bathtub), Zenith, Frigidaire, Winter Piano, arper Furniture, Norge, Acme Trucking, White orse Truck, Fruehauf Trailer and Thew-Lorain Shovel o. (Lorain, Ohio), to name a few. These hand-cast hite metal miniatures ("midgets," "novelties," and three-dimensional representatives") were produced s ashtrays, recipe boxes, working models, model rs and trucks, paperweights, medallions, cigarette ases, pin trays, desk ornaments, and numerous ther useful items. Some of the companies had the me item made as a bank and also as a papereight, model or desk ornament. These can be found both forms – with a coin slot and without a coin ot. These miniature metal models were given or sold at a very modest price to the customer as a constant reminder of the product. They also served as wonderful visual presentations for the salesperson to use in promoting and selling their product. The banks were especially useful in encouraging time-payment purchases of a refrigerator, car, etc. National Products and Banthrico used very similar sales promotion methods and ideas for their white metal miniatures which made it easy to merge them into one company.

People (Chapter 2)

The child on chamberpot is a difficult bank to locate, being both a bank and an ashtray. Pennies and dimes can be inserted in the open mouth. The Hopalong Cassidy bust is another color variation. The Seaman at the Wheel appears to be a Banthrico, but I am not absolutely positive; there are no company markings on the bank. Note that he has no beard or pipe and a different hat than the other seamen.

The General Washington bank (#934E) is a very early Banthrico bank with no name on the T1 trap. It was probably made in the 1930s. A picture of the T1 trap (#934F) is also shown since this particular trap (without company name) is not pictured in Chapter 12. It appears the early T1 Banthrico traps did not have the company name on them.

College and University Mascots and Sports (Chapter 3)

The plastic Baseball Cap and Hockey Puck banks were contracted out by Banthrico. They made one cap for each of the 24 American and National League Baseball Teams as well as hockey pucks for 43 teams in several hockey leagues and an association. There was no provision for customer advertising on the caps and pucks. They are all illustrated in Chapter 15.

Animals, Birds, and Fish (Chapter 5)

The Owl on the Limb bank has been elusive to locate. It appears to be made only for "The Michigan Bank," and thus has the "S" category. It is the same owl as the "Rice Owl" mascot bank (#100) with only a change in the lettering and the college seal. A flocked vinyl raccoon bank appears here also (#899).

Calculating, Registering and Pocket Banks and Coin Folders (Chapter 9)

Several additional cardboard coin folders have been included here to expand our knowledge of the generic and specific coin folders that were available. They do not appear on the flyer sheet.

Tranportation (Chapter 10)

In 1984, Banthrico decided to recast at least three promotional cars from their earlier molds. They were the 1954 Cadillac, 1950 Chevrolet, and 1957 Chevrolet and were advertised as "Bringing Back the Fifties." finished in an antique brass tone. Banthrico was expanding their product line to increase their sales and revenues. There was no expense in making the molds so this helped to keep down the cost of producing them. The Robt. E. Lee Steamboat is nicely done in a gold tone finish as a limited edition of 1000.

The Confederate Submarine is one of the "best finds" in this chapter. Only one submarine (#501) was known, now there are two! The "C.S.S. David" (Confederate States Ship) bank has "Oct. 5, 1863" and "Berkeley Bicentennial, 1976-1983" engraved on it. It is definitely a Banthrico.

The Kewanee Riveted Firebox Boiler bank was made for the Kewanee Boiler Company. It is a heavy white metal casting and a National Products item. It was also made as a paperweight and desk top item.

Architecture (Chapter 11)

The Great Northern Building and Loan Co. (#628) building is located in Barberton, Ohio. The picture of the actual building (#934C) and an architectural drawing of the building (#934D) confirm the 1955 date and automobiles of the 1950s in the architectural picture. The building is located at the corner of Third Street and West Park, but the Great Northern is no longer there.

Information

Charles M. Russell (#49) was born and raised in St. Louis, Missouri and moved to Montana as a young man. There is a place in University City (a suburb of St. Louis) called "The Loop" where famous people of St. Louis have a bronze star embedded in the sidewalk telling of their contributions. Charles Russell is among "the stars." It is something to see when you visit St. Louis.

A Herbert Hoover bust bank can be seen in *The Dictionary of Still Banks* by Earnest and Ida Long and Jane Pitman (#746). If it is a Banthrico it would be at least an "E" bank.

The Brahma Bull on Base (#129) is probably a bank made for Argentina or Columbia.

The name of the rocking horse (#181) is "Clyde" and the Purdue mascot#99 is "Pete."

Hopalong Cassidy (blue), Seaman at Wheel, Owl on Limb, Westinghouse Roaster, Toastmaster Toaster , Robt. E. Lee Steamboat (goldtone), and General Washington bust are all "D" banks.

Eight "E" banks that are even more difficult to locate are Child on Chamberpot, New York World's Fair Underwood Typewriter, Confederate Submarine, Kewanee Boiler, General Eisenhower bust (#928), Nash Ambassador, "Henry J.," and Corvette. The MacArthur bust (#924) is a very rare "F" bank.

The "Henry J." (#932) full-sized car was put on the market in 1951 by Kaiser-Frazer and lasted only a few years. The bank has a white metal and steel bottom, a plastic body, rubber tires and a white metal grille, lights and bumpers. Banthrico probably contracted out the plastic body and then installed the white metal and steel chassis and parts with the rubber tires on the plastic body. It was an autobank, a 1/25th scale model, and sold for $1.30 each in lots of 12-99, $1.12 each for lots of 1,000-up. It was also spring-motor powered and came in assorted colors. The full-size car sold for $1,299. An advertising flyer on the Henry J is in Chapter 15.

The Hattiesburg Federal Savings (#629) was issued earlier as the Hattiesburg Savings and Loan (#934B)(same building bank). It is another variation and there is no "Federal" in the earlier bank name.

The 1953-55 Chevrolet Corvette Convertible (#933) was a special run by Banthrico in 1974 for two gentlemen named Buchardt and Gatewood. It was 1/25th scale and came in three color combinations – red exterior/white interior, white exterior/red interior, and blue exterior/beige interior. It appears in the Corvette Collectibles newsletter dated 1976 and 950 were made in white and only 200 each were made of the red and blue models. Note the plastic windshield and the fancy hubcaps. A picture of the Corvette bottom (#934) shows the base plate, T6 trap, "#746" and "B/G" initials cast into the body in front of right rear wheel for the two gentlemen who commissioned the bank. There is also a paper stick-on instrument panel.

The bottom of the Santa Sleeping (#119) bank shows the name "Deno Burralli" preceded by " c " cast into the metal (#934A). He was probably the sculptor of the Santa bank. The name is only on some of the bank bottoms. This picture also shows a good example of the T6 trap with the slot in the back of the bank.

The Penny Bank Book by Andy and Susan Moore is a classic among still bank collectors. It quickly became "The Bank Bible" for still bank collectors everywhere. There are, however, some a eas in the book that need correcting regardin Banthrico banks, and Andy would be the first t welcome them.

Those that are not Banthrico: Davy Crockett wit Rifle (#332) and Shoe House (#1602);

Those that are Banthrico: Baseball Player (#22 Santa on Rooftop (#105), F.D. Rooseve (#145), Forlorn Dog (#408), and Map of U.S (#1609)

Those with wrong title or name: Safe Harbor Cha (should be Snug Harbor Chair) (#1374), 190 Baker Electric (should be 1900 Pillbox (#1526)

Those that are Banthrico as well as another com pany: Washington Bust (#118), Ben Frankli Bust (#122), Lincoln Bust (#129), and Lin coln Bust (#130).

All of the Banthrico banks discussed in this para graph were made in white metal. You will note othe companies made some of these banks in lead, a listed in *The Penny Bank Book*.

884. Child on Chamberpot
6 1/8", T42, G, c. 1930s, E
Also ashtray

885. Hopalong Cassidy
4 ¼", T35, G, c. 1950s, D
Blue Variation

886. Seaman at Wheel
5 ½", T6, S, c. 1940s, D
"Pilot Life Insurance Co."

87. Baseball Cap, Vinyl
3/8", T12, G, 1973, A
altimore Orioles (MD.)
merican League (AL)

888. Baseball Cap, Vinyl
2 3/8", T12, G, 1973, A
New York Yankees (NY)
AL

89. Baseball Cap, Vinyl
3/8", T12, G, 1973, A
akland Athletics (CA)
L

890. Baseball Cap, Vinyl
2 3/8", T12, G, 1973, A
Philadelphia Phillies (PA.)
National League (NL)

91. Baseball Cap, Vinyl
3/8", T12, G, 1973, A
ittsburgh Pirates (PA)
L

892. Baseball Cap, Vinyl
2 3/8", T12, G, 1973, A
St. Louis Cardinals (MO)
NL

893. Hockey Puck
5", T31, G, c. 1970s, A
St. Louis Blues (MO)

894. Hockey Puck – Back Side
NHL (National Hockey League)
Slot in Top. Vinyl

895. Panda Bear
5 ¾", T12, G, c. 1980s, B
Vinyl

896. Owl on Limb
5 ½", T1, S, c. 1950s, D
"The Michigan Bank"

897. Blue Pig with Flowers
3 ½", T12, G, c. 1970s, A
"Made in Japan" – Composition
"Japan" stamped on rear of pig

98. Happy Colorful Pig – 4 ¼", T26, G, c. 1960s, B
lue – Slot in Top
rst Fed Svgs - Pompano
each, Fl.

899. Raccoon
5 ¾", T12, G, c. 1980s, A
Flocked Vinyl

900. Norge Refrigerator
4 1/8", T6, S, c. 1930s, B
Slot in Back

01. Westinghouse Roaster
¼", T6, S, c. 1930s, D
lot in Back
"Westinghouse Automeal"

902. Magic Chef Stove
3 ½", T6, S, c. 1930s, C
Slot in Back
"Save with Magic Chef"

903. Toastmaster Toaster
2 ¾", T6, S, c. 1930s, D
Two Slots in Top

904. Umbrella on Globe
4 ½", T22, G, c. 1970s, C
Plastic – Yellow
Commercial Bank of Seattle, Wa.

905. R. C. Allen Cash Register
3", T6, S, c. 1940s, C
Slot in Back – Conversion

906. Typewriter
1 ¾", T6, G, c. 1930s, C
Slot in Top

907. Underwood Typewriter
1 3/8", T6, S, 1939, E
"New York World's Fair"
"1939"

908. Calculating Cylinder
2 5/8", T18, G, c. 1930s, C
Green Calculator
New York Life Ins. Co. "A few cents a day will protect your loved ones"

909. Church World Service
"Elkhart, Indiana"
6 ¼", T41, S, 1940s-80s, B
"Give a quarter a day"

910. First Federal Pig
6 ¼", T41, G, 1940s-80s, A
Dime Saver

911. The Squirrel Club
6 3/8", T41, G, 1977, A
"Stash Your Cash"

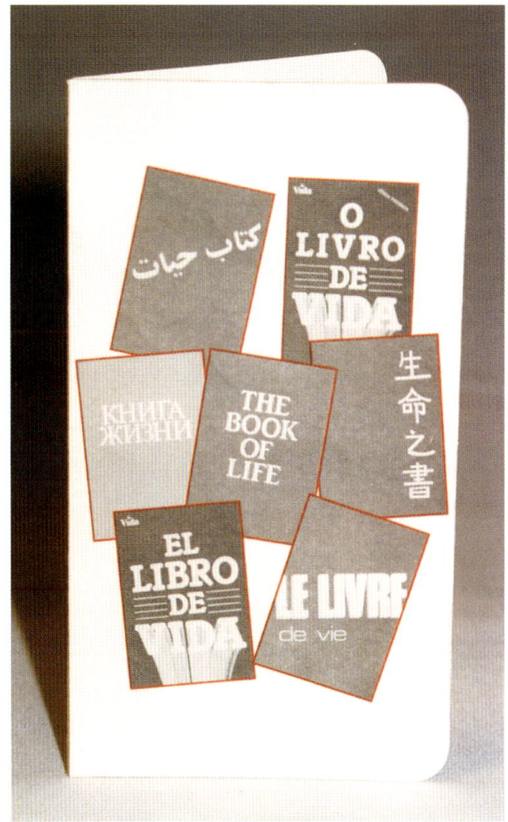

912. The Book of Life
6 ¼", T41, S, 1940s-80s, B
"Life Publishers International"

913. Christmas Stocking
"Hope Center – Miami, Fl."
6 ¾", T41, S, 1940s-80s, B

914. Heritage Olympia Bank
"Chicago Heights, Ill."
6 ¼", T41, G, 1940s-80s, A

915. Hoyne Savings & Loan
"Chicago – Since 1887"
6 ¼", T41, S, 1940s-80s, B

916. Cadillac – 1954
8 ¼", T10, G, 4, 1984, B
Promo Recast
Park National Bank

917. Cadillacs - 1954
L to R – The Old & New
Painted – Brass Tone

918. Chevrolet – 1950
7 ¾", T10, G, 4, 1984, B
Promo Recast

919. Chevrolets – 1950
L to R – The Old & New
Painted – Brass Tone

920. Chevrolet - 1957
7", T10, G, 4, 1984, B
Promo Recast

921. "Robt. E. Lee" Steamboat
7 ½", T5, G, c. 1970s, D
Gold Tone Lmtd. Ed.
First Fed. S&L Assn of East
Alton (Il)

22. Confederate Submarine
¾", T4, S, c. 1976, E
C. S. S. David"
'Banthrico Inc. Chicago USA"
1863-1983: 120 Years)

923. Kewanee Firebox Boiler
3 ¼", T6, S, c. 1930s, E
"Kewanee, Ill. "
"Riveted Boiler"

24. Gen. Douglas MacArthur
5/8", T43, S, c. 1942, F
Rose Santoro – R. S. ©1942"

925. General Eisenhower
5 ½", T3, G, c. 1946, D
Custom Painted – See #15
Painter: Tony Carmi

926. Baseball Player
6 ½", T12, G, c. 1980s, B
Vinyl, "Half Pints"

927. Welsh Rabbit
9 ½", T12, G, 1978, B
Vinyl Plastisols

928. General Eisenhower
5 ½", T3, G, c. 1946, E
Dress Blues/Gold Shoulder Boards
Note gold trim on hat/base
White Dress Shirt - Factory paint job. Same bank as #15 (Khaki uniform)

929. Chevrolet-Styleline Club
7 5/8", T25, G, 1, 1950, D
Two Toned Paint
Partial color sample/roof

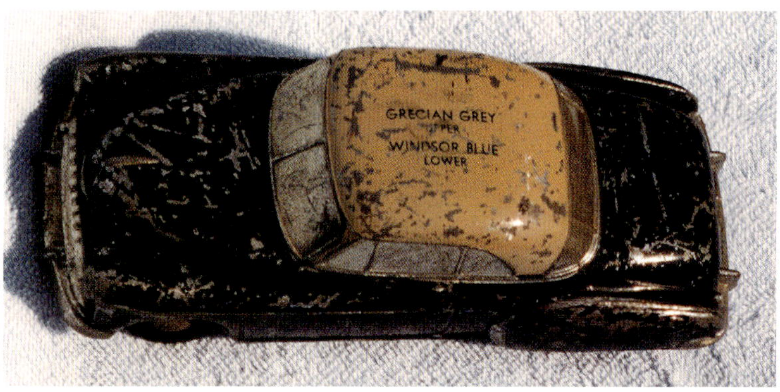

930. Chevrolet-Belair Hardtop
7 5/8", T25, G, 1, 1950, D
Show Room Color Sample
"Grecian Grey - Upper"
"Windsor Blue - Lower"

931. Nash-Ambassador
6 ¾", T25, G, 1, 1949, E
"Boise, Idaho"
"Bosse Nash Inc. - 601 Capitol Blvd."
Note "National Products" on tires
"Nash Airflyte" on trunk

932. Kaiser – Frazer's "Henry J"
7 ½", T25, G, 1, 1951, E
Plastic Promotional

933. Chevrolet Corvette – 1953-55
6 ½", T6, S, 1, 1974, E
Lmtd. Edition Convertible

934. Corvette Bottom
"B/G" Initials
"#746" of 950

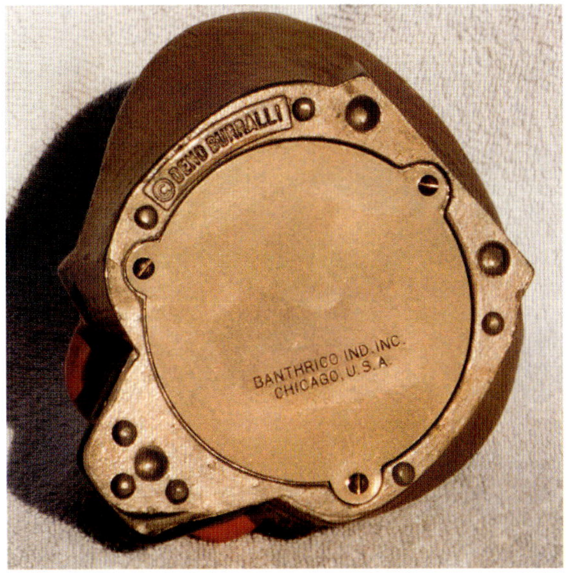

934A. Santa Sleeping (#119)
"© Deno Burralli"
T6 – "Banthrico Ind. Inc."

934B. Hattiesburg Savings & Loan (Hattiesburg, Ms.)
3 1/8", T4, S, 1960s, D

934C. Great Northern Bldg. & Loan Co. – Actual Bldg.
Barberton, Ohio

934D. Great Northern Bldg. & Loan Co. – Same as #628
Architectural Drawing

934E. General G. Washington
4 7/8", T1, G, 1930s, D
Military Uniform

934F. General G. Washington
T1 Trap – No Company
Name on Trap

934G. Building Bank Box
First Nat. Bank – Amarillo, Texas
with Key

934H. National Products
Construction Vehicles
Euclid/Terex
Note Bulldozer

934I. National Products
GMC/Sterling Trucks
Advertising Vehicles

934J. Actual Heights State Bank Bldg., Houston, TX

934K. Original Heights State Bank c. 1919

934L. Actual Beckley National Bank Bldg., Beckley, WV Now Bank 1 One

Chapter 14
Toystalgia and Valley Casting Company

The Valley Casting Company (VC), of Golden Valley, Minnesota, has a brief and interesting history. Darryl Knutson is the president of the company which is family owned by Knutson and his four adult children. Knutson was a CPA who joined Animal Fair, a company which made stuffed toys before starting his own company, Toystalgia, in January, 1976. He made wooden coin banks because he knew a lot of people who sold banks to financial institutions. Toystalgia's banks had elaborate burned decorations on them (#962). The Banthrico Company was sold in May of 1985 to Knutson, making Banthrico a division of Toystalgia. A box (#994) supports this fact. When Knutson purchased Banthrico he secured the total company and their production assets including 600 to 700 molds, the customer card file, and all their manufacturing machinery. VC uses the founding date 1931 on their advertisements which was Banthrico's start-up date. When VC purchased Banthrico, some of the molds or parts of molds were missing, having been destroyed or sold earlier as scrap by Banthrico. There are about 900 different banks known, which means about 200 molds were missing. Toystalgia did not receive Banthrico bank date and production records when they purchased the company.

Toystalgia stockholders started the company BankLand to make lithographed tin banks (similar to Chein and Ohio Art banks). BankLand was later liquidated by the stockholders. After that, Toystalgia purchased the ADCO Company of St. Paul, Minnesota, in February of 1988, which was a specialty advertising distributor that sold pens, calendars, etc. Later, the ADCO Company bought the Excaliber Company of Rhode Island which manufactured pins. In January of 1995, the Knutson family combined Toystalgia and the ADCO companies into Valley Casting, a new company for casting banks and pins. In November of 1999, Valley Casting purchased Showcase Specialties, which designs and manufacturers custom colored lapel pins.

In making their banks, Valley Casting used the steel bottom plates from Banthrico that have "Banthrico – 1974" on them. They also used Banthrico screws and plastic plug inserts until the supply was depleted. The new VC plastic plugs are bright and shiny pewter color (#1000) while the Banthrico plastic plug inserts are dark, dull gray. The plugs of both Banthrico and VC were/are made by the same supplier. The screws on the present VC banks are small Phillips screws, never seen on Banthrico banks (unless changed by someone). A date on a bank is another good indicator of its maker, after 1985 it is usually a VC bank.

While most of the white metal banks issued by VC were made originally by Banthrico, VC has added several banks of their own and they are included here. A listing "Cast Metal Coin Bank Stock Molds by Valley Casting" is shown in Chapter 15 (#1131). A price list (#1132) is also included, for ordering telephone 1-800-328-5765.

The majority of the VC white metal banks are painted and have a pewter finish for consistency of color and quality. Banthrico had a problem keeping their colors consistent and some were lighter than others. Banthrico had moved from the plating process to the painting process in the 1980s because of OSHA requirements. Good examples of Banthrico painted process banks are the 1950s Cadillac and two Chevrolets done in 1984 and discussed in Chapter 13. The pewter color remains the same for years later, which helps in matching banks like the train, tender, boxcar, and caboose. VC also uses other painted color finishes that can be special ordered. The Universal 1 Credit Union (#992) is a fine example of their brass tone painted finish. VC began using a shiny pewter finish, but switched to the dull pewter finish because it looked more expensive and had a higher perceived value; the shiny banks appeared cheap. A good example of a shiny bank is the Lincoln Cosmopolitan (#955). The white metal composition (zinc and aluminum) used by VC is the same as was used by Banthrico.

VC has a company in Chicago that makes their steel mold parts. They also use a sculptor to make the plaster shell for the mold company. For example, sculptor Lois Mihok made the Western Maryland Station Center and other banks for VC. She also did some sculpting work for Banthrico and she is located in Chicago. When the mold is completed, it goes to a tool and die maker to fit the mold parts together. The cost of an average mold is about six thousand dollars. The vinyl banks, designed by VC but contracted out, are rotation molded. The white metal banks are hand-cast by VC, which is a more labor-intensive process. (See Chapter 1 "Manufacturing.) VC also spin casts white metal pins to sell.

The VC production of white metal banks is being reduced and replaced by vinyl banks which are virtually indestructible. "The demand for metal banks is not there and the cost of producing them is getting higher and higher. The people who are selling the banks are more concerned about price than quality," observes Darryl Knutson.

Currently, VC offers a custom bank job in a minimum of one thousand pieces at $10 each. The mold start-up price of $6,000 plus engravings has to be provided by the customer. It is costly to have a new white metal bank made. VC takes the position that their custom mold rights are owned by the company ordering the banks and they will not reproduce them. Recently the Souvenir Building Network, headed by Dixie Trainer, contracted with Valley Casting to make two former white metal Banthrico banks — the Wrigley Building and The Exchange Bank of St. Augustine.

The mascot molds for colleges, universities, and a baseball team that were originally made by Banthrico and now are owned by VC have been used to make new banks for only a few customers in small numbers because colleges are getting very protective about their school mascots; they want a royalty on anything that has their mascot image. This practice raises the price of the bank and cuts into VC's profit margin.

Valley Casting has about 15 employees. They do not keep a record of the different banks they have made, the number they made, and the years they made them. Nor do they have records of different finishes or variations of the banks made. VC has two catalogs of the products they make. The 1990 catalog includes stock vinyl and white metal banks and white metal pins. The 1999 catalog is an update of the 1990 catalog which also covers stock vinyl and white metal banks and white metal pins and includes six custom building banks that Valley Casting has done. The catalogs emphasize that they can custom design banks from photos, blueprints, and drawings and will inscribe serial numbers on each custom-made bank. They sell stock banks to the public with a $100 minimum order. The catalog cover relates that "they have accumulated hundreds of stock molds since 1931 and may be able to meet your company's particular need." VC has issued a 2000-2001 catalog entitled the "Banksmith Collection" on cast coin banks. They have not kept an example of each bank they have produced. "Made in USA" has been added in the last three or four years to some of the VC bank bottoms.

Early lead banks, such as those made by Kronheimer and Oldenbusch, Rehberger, and a few Banthrico, had clear details in their moldings. Molten lead has a very fine viscosity and is quite smooth and works well in the molds so fine detailed banks with little or no casting marks could be made. Lead is an expensive metal and has to be handled carefully to avoid lead poisoning and conform to growing regulations. Banthrico switched to white metal (zinc and aluminum), a more economical metal for making their banks following World War II. The Banthrico banks in white metal have some very good detail, but not as good as the lead banks made earlier. More casting seams and marks are seen in the white metal banks than the lead ones.

The Valley Casting banks, in general, have less detail and more casting seams and marks than the Banthrico banks. Knutson recently stated that the ustomers today are more interested in price than uality, so that is taken into consideration when aking the white metal banks.

When rating the white metal Valley Casting anks, you have to take into consideration three acts. First, the recency of the bank issues — the ast 13 years (Western Maryland Station - 1986) ith the majority being made in the 1990s. Secndly, the banks have been issued mainly to buildg, bank and historical collectors and not to the ublic in general. More of the banks will be saved nd in better condition. Thirdly, there is a good hance that more of the same banks will be issued y Valley Casting as another edition. I have a flyer #1130) on the Western Maryland Station Center tating that 1,004 limited edition brass-tone finish anks were produced in 1986. I secured one in Aarch, 1995 (nine years later) that is a pewter one. I strongly suspect this is a second or third edition) issue. My bank has no number on the botom. The flyer is in the Flyer Chapter. That is one of he reasons for its rating of a "B." Therefore, the eneric banks are rated "A's" and the specific banks re rated "B's" and "C's." None are rated any higher han a "C." The ratings of the Valley Casting banks ill certainly have to be revisited in the next five to en years to see if these factors and ratings hold ue.

Related to the rating of Valley Casting banks the matter of Banthrico banks that have been eissued by VC, such as Northern Trust, Bangor avings Bank, the Wrigley Building, Exchange Bank f St. Augustine, Old Capitol Bank and Trust, Ruherford County Courthouse, Lincoln Cosmopolian, Fleetwood Cadillac, Ford Customline, and oths. These Banthrico banks are being recast by Valy Casting. The rating of a Banthrico Exchange ank of St. Augustine (#578) which was made in he 1950s is a "D," while the rating of the same Valley Casting bank (#997) made in 1999 is a "B." Chapter 1 "Pricing" can be consulted to determine the approximate price ranges.

Valley Casting has made popular original white metal banks since their first one in 1986 (#993). One is a bust (#935), two are mascots (#936 and #937), three are pigs, two are safes, thirteen are transportation pieces, and twenty-seven are buildings (nine are generic and eighteen are specific). There is also one vinyl building (#991). Banthrico never made a vinyl building bank. The two mascot banks are nice and fit well with the Banthrico mascot group. The John Deere Tractor series (6) makes a nice collection and goes well with the Banthrico tractor signs (6) found in Chapter 10. The boxcar and caboose banks complete the train set with Banthrico's train and wood tender. The Quaker is the first original bust by VC. The armored pig would be a good companion for the "Robo Cop" (the movie), and the golf cart is a nice gift for the golfer as well as the bank collector. The Schwan's Ice Cream Truck is a good addition for the transportation bank collector; it is not easy to find.

Among the original white metal building banks are ten "A" banks, eight "B" banks, and nine "C" banks. Brooke Mansion, Cape Ann Savings, Emigrant Savings Bank, Harford Mutual Insurance Co., McPherson County Courthouse, Mountain Springs Hotel, Universal 1 Credit Union and Western Maryland Station are "eye catching" and sought after by bank and building collectors. Several high-rise buildings are quite attractive, such as the Citizens Federal Centre, Continental Bank, and Norwest Indiana Center. The Gleaners Insurance Co. building is reminiscent of Trout Hall on the Banthrico side. The Harford Mutual Insurance Co. raises a question: is it the well-known Hartford Insurance Co. building with the "T" left out of the name, or is it actually Harford as spelled? The actual building photo (#984) of the Mountain Springs Hotel is also here.

The Statue of Liberty (#727) is dated 1986 and the bank was actually started by Banthrico and completed by Valley Casting, so both companies were involved in making this bank.

B&O Queen City Station (#535) was one of the first banks made by Valley Casting, according to Knutson, so it should be listed here instead of Chapter 11 and dated around 1986. The Western Maryland Station Center (#993) produced by Valley Casting is a companion piece to the B&O Queen City Station.

The Bangor Savings Bank was recast in 1992. It has "1852 to 1992" stamped on the bottom of the bank. It is a numbered edition of 1,000. The rest of the bank and information is the same as #537, but would be rated a "B." The Rutherford County Courthouse would also be rated a "B."

The VC banks are endlessly interesting. When you purchase a Seabees bank, about half of the bank cost goes to orphans and widows of disabled Seabees. The Northern Trust Co. bank (#985) back side is entirely different from the Banthrico issue, as shown in the picture. The Banthrico bank (#671) has a small anchor seal on the end of the building bank while the VC anchor seal is on the back and is much larger. The mini-van bank (#956) appears to be made in the style of the early wooden Toystalgia banks (#962). There are no sharp edges on the bank, just rounded parts (made with children in mind). Notice that two transportation banks have moveable plastic wheels (#944, #955).

The Reiter Dairy building banks (not shown) (#1-1995, #2-1996) resemble Banthrico's version and/or Valley Casting's. but they are neither; they are well done white metal banks executed by Antique (or Olde Mill) Designs Co. in Chardon, Ohio, 44024. There are no company markings on the banks.

935. Quaker
5 1/8", T10, S, 1998, B
"Quaker Oats Co."
Cedar Rapids, Iowa
"1873-1998 - Quasquicentennial"
Limited Edition - #1623

936. "MC" Bulldog on Stand
5 7/8", T8, S, 1987, B
"McPherson College"
McPherson, Kansas
"1887-1987 - Centennial"
"Affirming Our History - Assuring Our Heritage" Limited Ed. #0688
Done in brass tone also

937. U. S. N. Seabees
5", T6, S, 1992, B
Slot in Top
"Can Do - 50 Years - 1942-92"
Navy Seabees Mascot

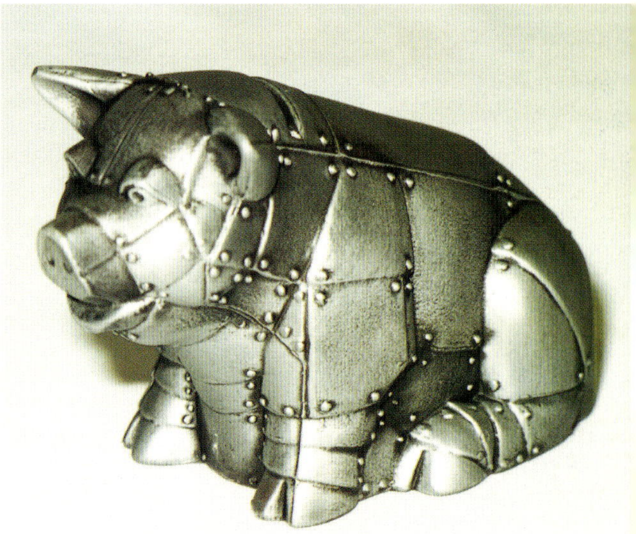

938. Armored Pig
3 ¾", T10, G, c. 1998, A
Slot in Top

39. "P. I. P. " Pig
½", T10, G, c. 1996, A
anthrico Mold

940. "R. B. Rice" Pig
3 ½", T10, G, c. 1995, A
Sausage Co.

941. Money Safe
4 7/8", T10, G, c. 1998, A
Slot in Top Also
Non-moveable molded dial

942. San Joaquin Safe
4 ¾", T10, G, c. 1998, A
Slot in Top Also
Combination dial molded into bank - does not move. Unusual black & gold bank. Similar to "Security Safe Deposit" bank #36-SBCCA Safe Bank Book

943. Cadillac – Fleetwood-1955
8", T8, G, 1, c. 1990s, B
Promo Recast

944. Chevrolet 3100 Pickup
7 ¾", T10, G, 5, c. 1990s, A
Note Fancy Hubcaps

945. Dodge Panel Truck – 1941-47
7 1/8", T10, S, 4, 1992, B
"Schwans Ice Cream"
"40th Anniversary 1952-1992" "Marvin"
on door "Marshall, Minn.-G.W. 7000"

946. Ford Panel Truck – 1934
6 1/8", T10, G, 4, 1992, B
Still Bank Collectors Club of America
1992 Lincoln Penny on bottom plate

947. Ford Customline - 1956
7 5/8", T8, G, 1, c. 1990s, B
Promo Recast

948. Golf Cart
5 5/8", T10, G, 4, c. 1990s, A
Golf Clubs in Bags

49. John Deere Tractor
½", T10, S, 2, 1990, C
irst Edition
"John Deere Community Credit Union -
Iodel A - 1934"

950. John Deere Tractor
6 ¾", T10, S, 2, 1992, B
Second Edition
"Waterloo Boy - Model N - 1916"

51. John Deere Tractor
7/8", T10, S, 2, 1993, B
hird Edition
"Model D - 1923"

952. John Deere Tractor
6 3/8", T10, S, 2, 1994, B
Fourth Edition
"1892 Froelich"

53. John Deere Tractor
¼", T10, S, 2, 1995, B
ifth Edition
"1928 Model GP"

954. John Deere Tractor
5 3/8", T10, S, 2, 1996, B
Sixth Edition
"1929 GP Wide Tread"

955. Lincoln Cosmopolitan – 1951
8 ½", T10, G, 5, c. 1990s, B
Silver Tone Finish

956. Mini Van
6 1/8", T10, G, 4, c. 1998, A
Wood Style Bank/ Toystalgia

957. Steamboat-Paddlewheeler
6 5/8", T10, G, c. 1995, A
Boatman's Bank Emblem. St. Louis, MO
Nickel plated plate. Purchased by
Nations Bank and then Bank of America

958. Steamboat on Water
8", T33, G, 1990, B
"Tall Stacks – 92"
"Port of Cincinnati - 1990 G.C.T.S.I."

959. Train Boxcar
6 1/8", T10, G, 2, c. 1990s, A
Black Also

960. Train Caboose
5 7/8", T10, G, 2, c. 1990s, A
Black Also

961. Truck – 1919
5 ½", T10, G, 3, c. 1995, A
"Alumax Foil Co."
"America's Oldest Foil Co."

962. Truck - Toystalgia
7 ¼", T12, G, 6, 1980, A
Wooden – "7-up"
"Golden Valley, Mn. 55427"

963. Badger State Bank
3 5/8", T10, G, c. 1988, A
(MN)

964. "Bank" Building
3 ½", T10, G, c. 1988, A
Stock Casting

965. The Bank of New Glarus (WI)
4", T10, S, 1993, B
Slot in Back Also
"1893 - 1993"

966. Brooke Mansion
(Lancaster, PA.)
3 ½", T10, S, c. 1998, C
No name on bank

967. Cape Ann Savings
(Glouster, MA.)
5", T10, S, 1996, C
"1846 - 1996"

968. "Church" Building
4 ¼", T10, G, c. 1988, A
Stock Casting

969. Citizens Federal Centre
(Dayton, OH)
5 3/8", T10, S, 1989, B

970. Continental Bank
(Chicago, IL)
5 ¼", T10, S, c. 1997, B

971. Continental Bank
On Base – "Est 1857"
Slot in Top Also
"The Future Holds a Lot of Change"

972. "Credit Union" Building
3 ½", T10, G, c. 1990, A
Stock Casting

973. Cumberland Insurance
3", T10, S, 1994, C
"1844 – 1994"
Slot in top also

974. East Penn Manufacturing
Bowers, PA.
3 5/8", T10, G, 1996, A
"50th Anniversary - 1946-1996"
"DELCO Batteries"

975. Easton National Bank
and Trust Co. (PA)
3 ¼", T10, S, c. 1990s, C
Art Metal Works also made earlier
bank. See MM, #871, p. 128

976. Emigrant Savings Bank
(New York City, N. Y.)
6 ½", T10, S, c. 1990s, C
Made in U.S.A." on bottom

977. Evansville Teachers
Federal Credit Union
3 ½", T10, G, c. 1990s, A

978. Federal Reserve Bank
of Dallas (TX)
3 ½", T10, S, c. 1990s, B
"1921" - Year building built. Slot
in top also. Note eagles over entrance.

979. First National Bank
of Milaca (MN)
2 5/8", T10, S, 1997, B
"1897 - 1997" Slot in roof
"#629 of 1000"

980. Gleaner Insurance Co.
3 7/8", T10, S, 1994, C
"1894-1994 – 100th"
"Limited Edition - #0079"

981. Har(t?)ford Mutual
Insurance Company
3 ½", T10, S, 1992, C
"1842-1992 - 150th Anniversary"

982. McPherson County Courthouse
(McPherson, KS.)
4 ¾", T10, S, 1994, C
"1894-1994"
"1894" over main entrance

983. Mountain Springs Hotel
(Ephrata, PA)
4 3/8", T10, S, 1991, C
"Centennial Year of the Borough of
Ephrata 1891-1991" "First Edition - #0178".
Hotel name on bottom of bank

984. Mountain Springs Hotel
Actual Building
Built in 1840

985. Northern Trust Co.
(Chicago, IL) – Back Side
4", T10, S, 1989, A
"A Century of Service - 1889-1989"
"Principles that Endure"
Slot in top and bottom

986. Norwest Indiana Center
(Fort Wayne, IN.)
4 ¼", T10, S, 1995, B
"Est. 1995" Slot in roof also.
Bronze also.

987. Osceola County Courthouse
(Kissimmee, FL)
5 5/8", T10, S, 1990, B

988. "Post Office" Building
3 ½", T10, G, c. 1990s, A
Stock Bank

989. Post Office
Decatur, AL – 35602
3 ½", T10, G, c. 1990s, A

990. "School" Building
4 ¼", T10, G, c. 1990s, A
Stock Casting

991. Two Story House
First Vinyl Bldg. In Book
3 3/8", T11, G, c. 1996, A

992. Universal 1 Credit Union
(Beaver Creek, OH.)
2 1/8", T10, S, c. 1997, C
World globe on front

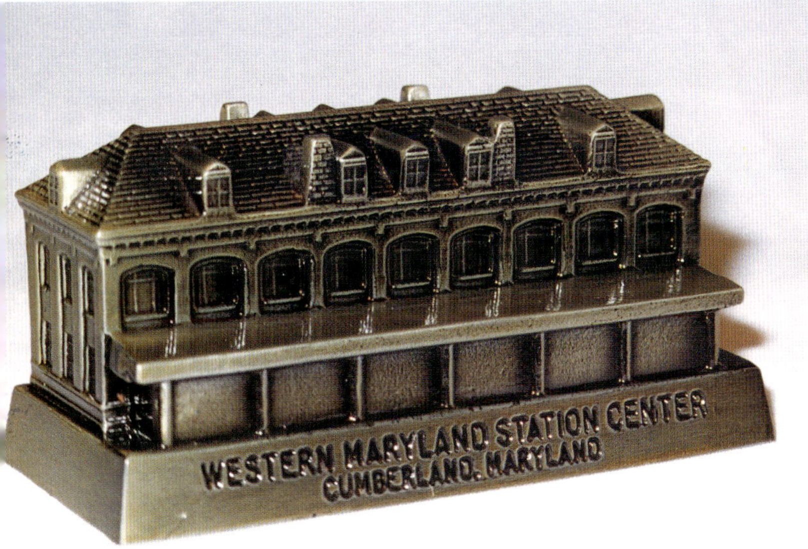

993. Western Maryland Station
Center – Cumberland, Maryland
3", T8, S, 1986, B
Chessie Employees Fed. C.U.

994. Banthrico – A Division
of Toystalgia – 1985
Box – End View
"Golden Valley, Minnesota 55427"

995. Banthrico Box – Side
Same Box as #994
Note 1931 Date

996. Continental Bank
in Presentation Box
Limited Number

997. Exchange Bank
of St. Augustine (FL)
5 ½", T10, S, 1999, B

998. Wrigley Building/
Boulevard Bank
7 ½", T10, S, 1999, B

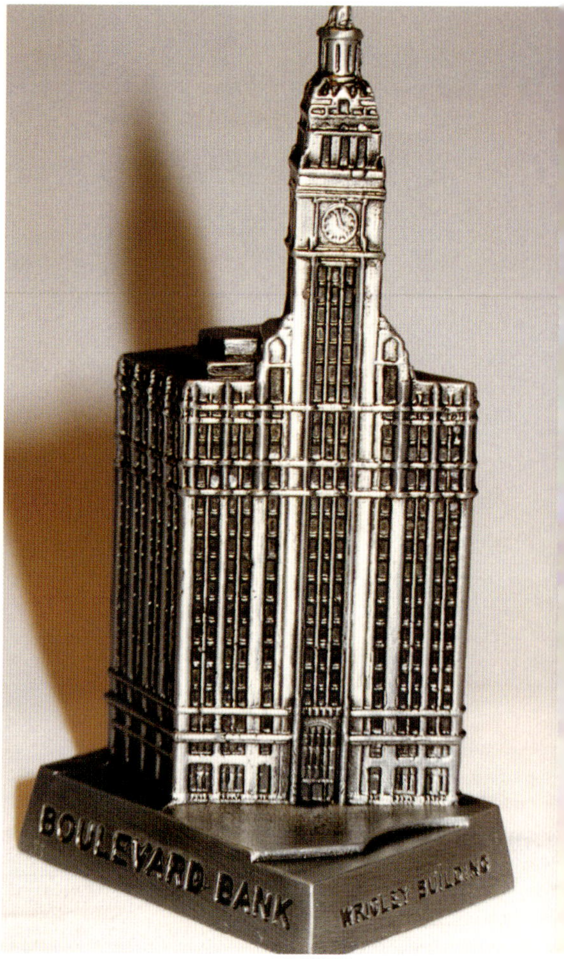

999. Osceola County Courthouse
"100 Years of Justice – 1890-1990"
Pamphlet with Bank – 69 pp.

1000. T10 Trap – VC Insert
Shiny Pewter Plastic
Norwest Indiana Center (#986)
Note Phillips Screws and
"1974 Banthrico" in base plate.
"Made in U.S.A."/back

1001. McPherson Opera
House (McPherson, KS.)
3 ¼", T10, S, 1999, B
"1888 - #640/1000"

Chapter 15
Advertising Flyers and Letters

Advertising flyers and letters that Banthrico, Stronghart, Bankers Thrift, National Products and Valley Casting produced are represented in this chapter. Only a few examples are provided of the latter four companies. The majority of black and white flyers are not included here because they were replaced by colored flyers. Throughout the book reference has been made to advertising flyers in this chapter. The most important ones have been included.

Several of the banks shown in the flyers are not shown in the earlier chapters. There is a beautiful "Gemini Space Craft Bank" (#1096) that no one seems to have and it appears to be metal. It was made at the Des Plaines Street location so it was probably metal since this was made in the 1960s. McDonald Aircraft Co. of St. Louis is mentioned in the flyer, but the bank is unknown otherwise. It also has a trap that has not been listed in the book. Several of the building banks shown in the flyers also are not in Chapter 11. One good example is the National Bank of Wilson (#1104). The pictures are in the flyers, but where are the banks to match the pictures?

A second major factor about the flyers is that many of the banks in the picture catalog do not have flyers showing them. Flyers on the banks may have been printed, but perhaps all have been destroyed. There are no flyers on such great bust banks as Thomas Dewey, Dowagiac Indian, Alexander Hamilton, Theodore Roosevelt, F.D. Roosevelt, Will Rogers and Grant Wood to name just a few. There are only a few flyers on the mascot banks: Iowa State Hurricanes; St. Louis Cardinals; Alabama Crimson Tide; and Minnesota Gophers. Many of the mascots are not included. The submarine banks (2) are not shown. There are only a few trophy banks illustrated.

When examining the flyers, notice the different dates on some of the flyers or events that allude to the date. The majority of flyers in color were completed at the Roosevelt Road (1970-1985) location due to advances in color printing. The transportation section is by far the largest representation.

This chapter is organized as nearly as possible by the arrangement of chapters in this book: There are some display and price list flyers near the end of the chapter that show the gradual increase in the cost of the banks as the years advanced.

Valley Casting flyers at the end of the chapter include one page lifted out of their 1999 catalogue (which can be ordered from them if interested).

You will note in Chapter 11 a picture of the First Federal Savings and Loan Association of Detroit (#592A, #592B), and in this chapter is a picture of a building (#1098) with an addition behind it. The Olde Discount Corp. in Detroit recently sent me a pamphlet of their headquarters. Lo and behold it is the old First Federal Savings and Loan Association of Detroit building! Look closely and you will see the building in front and the Griswold Street address.

There is a good representation of building banks in the flyers. One building has two different names on it. First it is listed as the "Straus Building - Chicago" (#1097)and in another flyer it is listed as the "Continental Assurance Building" (#1098). The only difference detectable in the two buildings is the front door entrance and front windows. The building was probably renamed from Straus to Continental. *Monumental Miniatures* lists the Straus Building (#1177) as an A.C. Rehberger bank. It was also a Banthrico bank, which supports my theory that Banthrico and Rehberger shared or exchanged molds and information since they were both in Chicago. It appears that the Banthrico Continental Assurance Building (lead) (#565) was made in the late 1930s while the Rehberger Straus Building (lead) was executed earlier in the 1920s. The Banthrico Straus Building was probably molded in the mid-1930s.

In the historical section of the flyers, notice the Banthrico plant located on Roosevelt Road which accommodated 60 workers plus 15 salesmen. Many of the workers' cars are parked on the roof. Also notice the picture in this section showing some wooden boxes ready for shipment entitled "25 thousand Banthrico banks for Bogota, Columbia" (#1004). If you ever go to Columbia, Mexico, Puerto Rico, or Argentina, check the antique shops for Banthrico banks. The Banco Altantida Building (#536) from Mexico and the First Federal Savings and Loan Association of Puerto Rico (#597) are both "F" banks.

The first Banthrico sales manual is interesting reading and clearly shows the early stages of the company in merchandising their banks and other items (#1005). The Banthrico Inc. letter dated 1934 is the original signature of the first president, Jerry Aronson (#1007). Notice the simple telephone numbers and list of predecessors on the letterhead. This was at their first location – Lake Street in the 1930s and they state that they were "manufacturers of savings banks" and several other items. Also notice Ben Franklin which was their early company logo. The letter tells of Banthrico buying out a sign manufacturer. It would be interesting to know how many companies Banthrico purchased during their history.

National Products, Inc. is one of the five known companies which was purchased by Banthrico (1940). There are six flyers and letters that provide considerable insight into the company's operation. Even as late as 1947, the National Products name was still being used and the numerous small white metal advertising banks they made still had "National Products – Chicago" on the bottom of them. In addition, they made the very popular larger white metal scale model cars before the Banthrico buyout which became banks in the mid-1930s. They made some advertising pieces (banks and other items) and many were continued by Banthrico after they purchased them. In 1948-1949, the Banthrico name began to appear on the National banks and many of them were phased out. Banthrico was making steel, lead, and white metal buildings prior to purchasing National Products. The other National Products banks are discussed at length in Chapter 13.

The flyer on "Toy Town," Winchendon, Massachusetts, is located in this chapter since it involves a rocking horse (#1041). Banthrico made a rocking horse bank (#181) for "Toy Town" in the 1950s and it is rated an "E" bank. "Toy Town" became famous in the early 20th century as the home of Converse, the nation's largest wooden toy factory which was in business until the mid-1940s.

The 1978 Price List (#1124) and several other following price lists show a "Kansas Herky" mascot bank that is otherwise unknown. Another bank with new flyer information is the "Snug Harbor" Chair (#178). It was first a National Products bank and made for the Karpen Furniture Company.

1002. Banthrico advertisement flyers folder/front cover

Banthrico . . . Exclusively serving the financial community

Dynamic, customer-attracting promotions have been our only business for more than fifty years.

Offering the largest selection of unusual and unique savings devices, most financial institutions around the world rely on Banthrico to attract new accounts and keep established ones coming back.

Banthrico sells on a factory-to-user basis, eliminating costly jobber and middleman costs. Your satisfaction is our greatest concern. To assure this we offer you the complete services of our creative staff and art department in tailoring a promotion to fit your particular needs at no obligation. We furnish a complete selection of advertising and display materials at no cost.

Next time you think of a savings promotion, think of Banthrico . . . the coin bank people.

Banthrico executives (from left): Thomas Aronson, VP-sales; Joseph Eisendrath, president and cofounder; Peter Eisendrath, executive VP and treasurer. For 50 years the company has been creating coin banks for bank sales promotion.

APRIL 1982

1003. Banthrico management and building

1004. Banthrico banks for Columbia, So. America

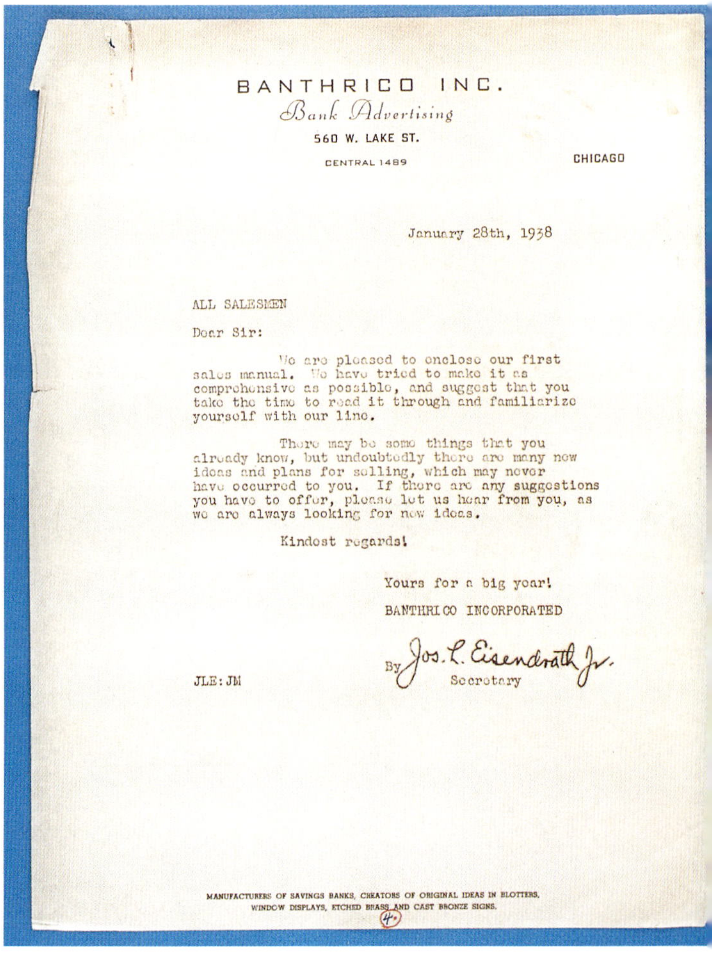

BANTHRICO INC.
Bank Advertising
560 W. LAKE ST.
CENTRAL 1489 CHICAGO

January 28th, 1938

ALL SALESMEN

Dear Sir:

We are pleased to enclose our first sales manual. We have tried to make it as comprehensive as possible, and suggest that you take the time to read it through and familiarize yourself with our line.

There may be some things that you already know, but undoubtedly there are many new ideas and plans for selling, which may never have occurred to you. If there are any suggestions you have to offer, please let us hear from you, as we are always looking for new ideas.

Kindest regards!

Yours for a big year!
BANTHRICO INCORPORATED

By Jos. L. Eisendrath Jr.
Secretary

JLE:JM

MANUFACTURERS OF SAVINGS BANKS, CREATORS OF ORIGINAL IDEAS IN BLOTTERS, WINDOW DISPLAYS, ETCHED BRASS AND CAST BRONZE SIGNS.

1005. Letter about first Banthrico sales manual

BANTHRICO INCORPORATED

GENERAL INFORMATION AND INSTRUCTIONS FOR ALL SALESMEN

The first thing we suggest you do is to take out your catalogue, and follow through this instruction list page by page. It will facilitate your becoming more familiar with the line.

COMMISSION — First of all, your commission -- This is paid at the rate of 20% for Savings Banks and Cast Bronze Signs, and 25% for Etched Brass Signs and Blotters. Our usual way of paying commission is to send you a remittance for one half the commission when the order is accepted by us, and the balance when the invoice has been paid to us. If, on orders for split shipments, as you will find explained later on, there is commission due, part will be paid only as each shipment goes out; in other words, if you have an order for future delivery, your commission for the shipment at that time will be paid at that time.

PRICES — From time to time you will note that we are compelled to change our prices, due to the cost of material increases or decreases. In such events, we will send you new catalogue sheets for your catalogue, and you can easily replace them.

TERMS — Inasmuch as the greater part of our cost is for labor in assembling the banks, and commission which is paid to you, you can appreciate that it is not possible for us to grant our customers cash discounts. Therefore, we have set our terms at net cash, due in 10 days after shipment of the merchandise. If your customer is a mercantile concern, and is not rated by Dun and Bradstreet, obtain credit references, preferably in Chicago. The customer always pays the charges for transportation, which, of course, are F. O. B. factory. Unless we have specific instructions as to the manner of shipment, we use our own discretion. Small orders usually go forward by express, and larger orders by freight or motor truck. It all depends on how quickly your customer wants the merchandise, as to which mode of transportation is adopted.

ORDERS — As you will note on the order blank, all merchandise is specially made for each order,

(5)

1006. Sales manual information

TELEPHONE CENTRAL 1489

BANTHRICO INC.

SUCCESSORS TO
BANKERS THRIFT CORP. AND STRONGHART CO.

EXPONENTS OF THRIFT

November 22, 1934

Central Company
Minneapolis, Minn.

Gentlemen:

Enclosed you will find a circular describing the most beautiful sign ever offered to the banking institutions of the United States.

You will note that with each sign, there are included four beautiful slides, the sign and slides having a retail value of $23.50. We are offering them to you complete at only $10.00 each.

The manufacturer of this beautiful sign has been forced to retire from business and we have taken over his entire stock of signs at a very low price and as a consequence we in turn are able to offer these beautiful signs to a few of our friends at less than half of what they are really worth.

There are only 55 signs available; orders will be filled in the rotation that they are received. First come, first served. Send in your order at once for as many as you desire. Order blank and return envelope are enclosed for your convenience. Price of sign with four changes of copy complete $10.00 F. O. B. Chicago.

Yours very truly,
BANTHRICO INCORPORATED
By J B Aronson
President

JBA:DH

560 W. LAKE STREET · CHICAGO · U.S.A.

MANUFACTURERS OF SAVINGS BANKS, CREATORS OF ORIGINAL IDEAS IN BLOTTERS, WINDOW DISPLAYS, ETCHED BRASS AND CAST BRONZE SIGNS, BUSINESS EXTENSION. (12)

1007. 1934 letter signed by Jerry Aronson

CIRCA 1947

NATIONAL PRODUCTS, INC., MAKERS OF MINIATURES IN METAL

This 4-page letterhead is genuine pre-war stock; with a paper shortage still on, we hope you won't censure us for using it.

17 N. DESPLAINES ST.
~~564 W. LAKE ST.~~
CHICAGO, 6 ILL.

Gentlemen:

Metal castings are again "in the picture"--and we're turning them out by the hundreds daily. Interesting little things they are, too: miniatures of the post-war products of several leading manufacturers who have learned their advertising value by extensive past experience.

There's something about a miniature--of almost anything--that's fascinating. "Shrink" a product of YOURS, new or old, down to a fraction of its normal size...give it the color as well as the accurate form of the life-size original...and you have a 3-dimensional representative that's as high in appeal as it is rich in distinction.

Suppose you let us figure on something like this for you. A product, trade-mark or trade-character in the form, let's say, of a paper-weight or ash-tray would make your presentation functional and increase its value to the user--to you as well.

These hand-cast metal novelties serve as enduring reminders of your business on the DESKS OF YOUR CUSTOMERS. They keep right on paying dividends--at trifling cost per year.

May we supply more specific information? The obligation will be all on our part for any such opportunity you give us.

Sincerely yours,
Norman Meyer
Director of Sales

PURCHASED BY
BANTHRICO IN 1940

(20)

1008. 1947 National Products advertisement letter

1009. National Products banks, paperweights, ashtrays, and desk items

THESE DAYS you want your advertising to be geared . . . and in high . . . to

(1) promote your products and services so as to step up sales TODAY and/or

(2) build for the future without sacrifice of present goodwill—to hold a place in today's market and create an acceptance for tomorrow, after the tumult dies and America has truly become the arsenal of Democracy.

As a valuable means to the end of meeting these vital needs, we present Miniatures in Metal-your product or trade character "shrunk" to scale, hand cast in indestructible alloy.

These provide as closely as it's possible to get

PERPETUAL proMOTION

— effective, inexpensive and above all *enduring* reminders of your business on the desks of the key men you want most to reach.

Perhaps what you see here will suggest a way in which you can make profitable use of a miniature. Or possibly you would prefer us to suggest a special idea adapted to your particular needs. We'll do it gladly.

NATIONAL PRODUCTS, Inc.
564 West Lake Street, Chicago, U. S. A.

THE MINIATURES ILLUSTRATED

1. Minute Man paperweight and desk ornament, 4¾" high, finished in statuary bronze. 2. Miniature building (a bank). Plants, office buildings may be "shrunk" to scale for use as paperweights. 3. Kewanee Riveted Firebox Boiler, a paperweight that has been long and successfully distributed. 4. Medallion paperweight, produced for Metro-Goldwyn-Mayer. 5. Miniature Burgess Snubber (a muffler) done for the Acoustic Division of Burgess Battery Co. 6. New-type gasoline pump, developed as both bank and demonstrating model. 7. Miniature Sanitizor Vacuum Cleaner—a coin-bank that aids in time-payment sales. 8. Norge Refrigerator miniature, also a bank. 9. Another bank—an Estate gas range. 10. Handsome ash tray for Delco Motors. 11. Novel ash tray produced for Hotpoint, with removable "grate" top in counterpart of the famous Calrod electric heating unit. 12. A miniature demonstrator of the White Horse Truck, with removable rear-wheel-attached power unit. 13. Stake Truck miniature for White Motor Company. 14. Buick found this extremely valuable during the 1941 model year. 15. A paperweight Fruehauf Trailer, designed for distribution by Fruehauf customers. 16. Westinghouse Electric Roaster in miniature—a bank. 17 & 18. Typewriter replica banks for Underwood and Remington-Rand (we've also made them for Royal).

1010. National Products vehicles and banks

Perpetual PROMOTION

is here today—

and not gone tomorrow, either!

That's one big advantage of metal miniatures. But *only* one. Consider, if you please—

- something more closely related to your business than your mother-in-law is to you . . .
- something that makes yawning wastebaskets snarl in frustrated rage . . .
- something that, on a cost-*per-year* basis, will cause even *your* comptroller to smile complacently!

Isn't this enough to bring a sparkle to the eye of any sales-minded executive . . . to yours, for example? Isn't it enough to justify your being classified at least as an "interested prospect"?

Do follow up . . . today. Tell us the type of item on which you'd like specific information and an estimate as to cost. Or direct our thinking along the general line of yours, so that we may submit, for your leisurely consideration, suggestions which you might label both practicable and valuable. No semblance of obligation; we'll be indebted to you for the opportunity thus granted to put the full story in your hands.

NATIONAL PRODUCTS, Inc.
564 W. LAKE ST. CHICAGO

p. s. *Here's a heartfelt, if slightly premature, Thank You for including with your request for further facts illustrations of the item you might want to make a "midget" out of. And will you please try to remember to give us the approximate size you have in mind for the completed miniature?*

21 "Sales-building" Uses

The producers of toy replicas mentioned are mainly toy manufacturers, and the miniatures they make are designed primarily as playthings. It is interesting to study the other side of the picture, a company that specializes in producing replicas as advertising specialties. National Products Corp., Chicago, has produced dozens of replicas for such well-known firms as International Harvester, Diamond-T Motor Car Co., White Motor Car Co., Reo and Studebaker (all trucks); and automobiles for Buick, Studebaker, Chrysler and Reo. Its illustrated circular shows miniature reproductions of other products, with illuminating comments on their use by the companies represented.

Other miniatures shown include: Westinghouse air conditioning equipment in the form of an ash-tray; an Ingersoll barrel in the form of a paper-weight; "Cleenie," the A.C. Spark Plug (the laughing horse in a bathtub)—sent to 60,000 dealers; coin banks that are models of Zenith radio, Frigidaire, Winter piano, Karpen chair, Norge refrigerator, etc.

The circular lists 21 "sales-building" uses for replicas, among which are: Door openers for sales canvassers; desk souvenirs; reminders (for installment buyers); "give-aways" to prospects and their children. Purposes mentioned include: To obtain names for dealer follow-up; to reduce substitution; to give sales force talking point; to insure better product display.

from SALES MANAGEMENT MAGAZINE

1011. National Products advertisement flyer

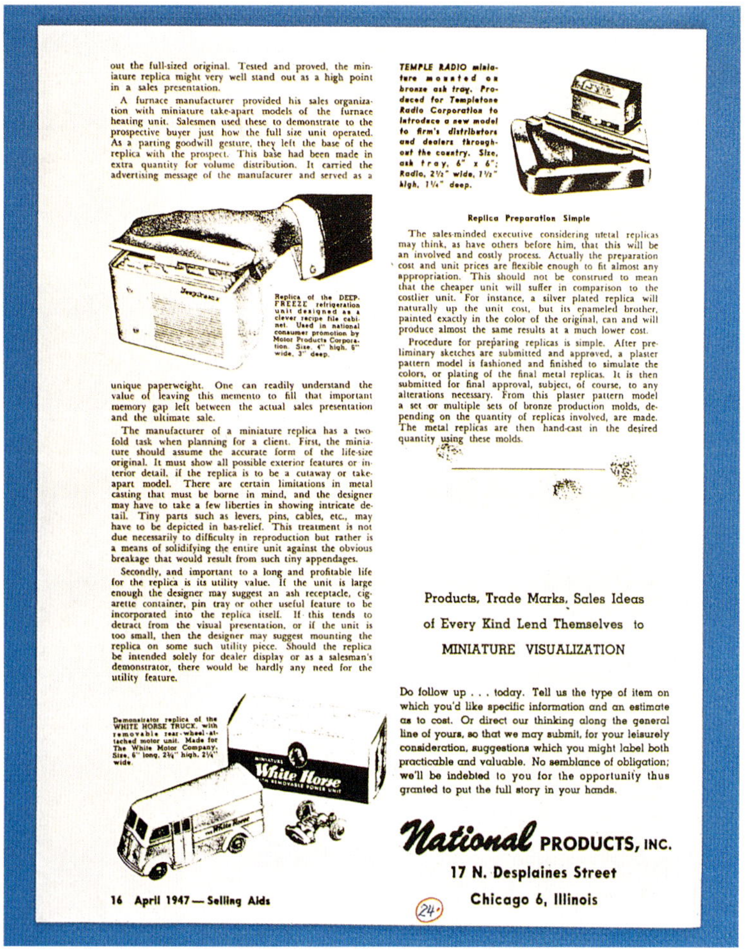

out the full-sized original. Tested and proved, the miniature replica might very well stand out as a high point in a sales presentation.

A furnace manufacturer provided his sales organization with miniature take-apart models of the furnace heating unit. Salesmen used these to demonstrate to the prospective buyer just how the full size unit operated. As a parting goodwill gesture, they left the base of the replica with the prospect. This base had been made in extra quantity for volume distribution. It carried the advertising message of the manufacurer and served as a unique paperweight. One can readily understand the value of leaving this memento to fill that important memory gap left between the actual sales presentation and the ultimate sale.

Replica of the DEEP-FREEZE refrigeration unit designed as a clever recipe file cabinet. Used in national consumer promotion by Motor Products Corporation. Size, 4" high, 6" wide, 3" deep.

The manufacturer of a miniature replica has a twofold task when planning for a client. First, the miniature should assume the accurate form of the life-size original. It must show all possible exterior features or interior detail, if the replica is to be a cutaway or take-apart model. There are certain limitations in metal casting that must be borne in mind, and the designer may have to take a few liberties in showing intricate detail. Tiny parts such as levers, pins, cables, etc., may have to be depicted in bas-relief. This treatment is not due necessarily to difficulty in reproduction but rather is a means of solidifying the entire unit against the obvious breakage that would result from such tiny appendages.

Secondly, and important to a long and profitable life for the replica is its utility value. If the unit is large enough the designer may suggest an ash receptacle, cigarette container, pin tray or other useful feature to be incorporated into the replica itself. If this tends to detract from the visual presentation, or if the unit is too small, then the designer may suggest mounting the replica on some such utility piece. Should the replica be intended solely for dealer display or as a salesman's demonstrator, there would be hardly any need for the utility feature.

Demonstrator replica of the WHITE HORSE TRUCK, with removable rear-wheel-attached motor unit. Made for The White Motor Company. Size, 6" long, 2¾" high, 2¼" wide.

TEMPLE RADIO miniature mounted on bronze ash tray. Produced for Templetone Radio Corporation to introduce a new model to firm's distributors and dealers throughout the country. Size, ash tray, 6" x 6"; Radio, 2½" wide, 1½" high, 1¼" deep.

Replica Preparation Simple

The sales-minded executive considering metal replicas may think, as have others before him, that this will be an involved and costly process. Actually the preparation cost and unit prices are flexible enough to fit almost any appropriation. This should not be construed to mean that the cheaper unit will suffer in comparison to the costlier unit. For instance, a silver plated replica will naturally up the unit cost, but its enameled brother, painted exactly in the color of the original, can and will produce almost the same results at a much lower cost.

Procedure for preparing replicas is simple. After preliminary sketches are submitted and approved, a plaster pattern model is fashioned and finished to simulate the colors, or plating of the final metal replicas. It is then submitted for final approval, subject, of course, to any alterations necessary. From this plaster pattern model a set or multiple sets of bronze production molds, depending on the quantity of replicas involved, are made. The metal replicas are then hand-cast in the desired quantity using these molds.

Products, Trade Marks, Sales Ideas of Every Kind Lend Themselves to MINIATURE VISUALIZATION

Do follow up . . . today. Tell us the type of item on which you'd like specific information and an estimate as to cost. Or direct our thinking along the general line of yours, so that we may submit, for your leisurely consideration, suggestions which you might label both practicable and valuable. No semblance of obligation; we'll be indebted to you for the opportunity thus granted to put the full story in your hands.

National PRODUCTS, INC.
17 N. Desplaines Street
Chicago 6, Illinois

16 April 1947—Selling Aids

24

1012. National Products information bulletin

from Selling Aids

Miniature THEW-LORAIN crane developed for The Thew Shovel Company as both a demonstrating model and unique paperweight. Size, 5½" long, 3" high, 3" wide.

MINIATURE REPLICAS Visual Selling Aids

By J. Bernard Strauss, *Director of Advertising & Sales, National Products, Inc., Chicago*

METAL CASTING in miniature, Lilliputian in name only, is becoming an important factor to consider in well planned advertising and sales campaigns. It can play a useful role in permanently establishing the product and sales story in the user's mind. There's something about a miniature that's fascinating to many people. Look in on Bill Smith in the next office. He's probably toying with some miniature gadget or flicking his ashes into a replica which has been cannily designed for that purpose. Each one of these had been given him on the very assumption that he would do just that. Whether or not he realizes this, a keen awareness of the product is being formulated in Bill Smith's mind. Possibly he will have a strong feeling of loyalty for a concern who has provided him with this attractive piece. Of course, there's no obligation on his part, but human nature and good nature conspire within him to induce an ultimate purchase.

Use in Dealer-Consumer Relations

This holds true in dealer-consumer relations as well. The replica coin-bank is one example of consumer promotion being successfully used to sell practically any merchandise. The manufacturer, in cooperation with his dealers, has a quantity of replicas of his product made in the form of a coin bank. These are distributed to the dealer who, in turn, may give the bank away as a premium with the sale of the product. Another successful method of coin bank replica use is to encourage time-payment purchases. The dealer hands out the bank to every prospect, charging a small sum which is returned when the bank is brought in and its contents used as an initial payment on the product. Therefore the bank belongs to the customer and almost invariably it is used to save for periodical payments.

The coin bank replica is only one example of the application of replicas. To cite a few examples: A large motor car corporation distributes through its dealers a miniature replica of the new model to every anxious purchaser on their dealers' "lists," thus retaining goodwill and building for future sales. The dealer sends a miniature replica with a personal letter to each buyer which reads in part—"and we want to thank you for your valued order which, due to present conditions, we cannot fill as yet. Please accept this miniature as a token of our appreciation for your complete confidence and understanding. You may feel sure of receiving its big brother just as soon as we can make it available to you." Instead of various excuses and promises, which usually put the buyer in a negative state of mind, a replica of the forthcoming model did much to inspire his customers with new confidence and an extreme feeling of loyalty.

A manufacturer of major electrical appliances used a miniature replica of a freezer unit to introduce his product to a highly competitive market. The replica was made in the form of a recipe file box and included a set of card separators into which the housewife could file various recipes. To date, this promotion has been successful in establishing the manufacturer's product.

It is worthy of passing comment that no one use of miniatures can be aimed at without securing incidental profit in another direction. Consumer sales can accrue to a company as a result of say, a replica intended solely as a goodwill item. In one instance a major motion picture corporation had decided upon a replica of its trade character in the form of a simple paperweight as a twentieth anniversary memento to be given to distributors and exhibitors throughout the country. The popularity of this replica paid unusual dividends. In a short time theater owners were ordering these in quantity to use as give-away premiums for bolstering attendance. In some cases theater owners made a display of replicas in their lobbies and sold these as souvenirs. This occurrence is not unusual. Time after time the consumer discovers unthoughtof uses for replicas which in the analysis, add up to more sales and goodwill for the user of replicas.

Demonstrating the Product in the Flesh

The miniature replica is a versatile little fellow, carrying through in many directions and for a wide variety of uses. For instance, if a product is too unwieldy for a salesman to take conveniently or advantageously to the buyer, as in the obvious case of heavy machinery or equipment, the replica maker makes a good proxy. The customer or prospect can obtain a true visual idea of the hard-to-see features ordinarily impossible to show with-

Take-apart model of the TUBERATOR heating unit done for Wheeling Furnace Corporation. Assembled quickly to show working parts, this unit has been proved a boon to the Wheeling sales organization. Size, 5½" high, 4½" wide.

April 1947—Selling Aids 15

25

1013. Miniature replicas by National Products

014. Famous American bust banks

1015. Davy Crockett bust bank

016. Banthrico banks of the 1970s and 1980s

1017. Banthrico banks of the 1950s and 1960s

1018. Plastic mechanical clown bank

1019. Hopalong Cassidy Savings Club flyer

1020. Hopalong Cassidy bust bank

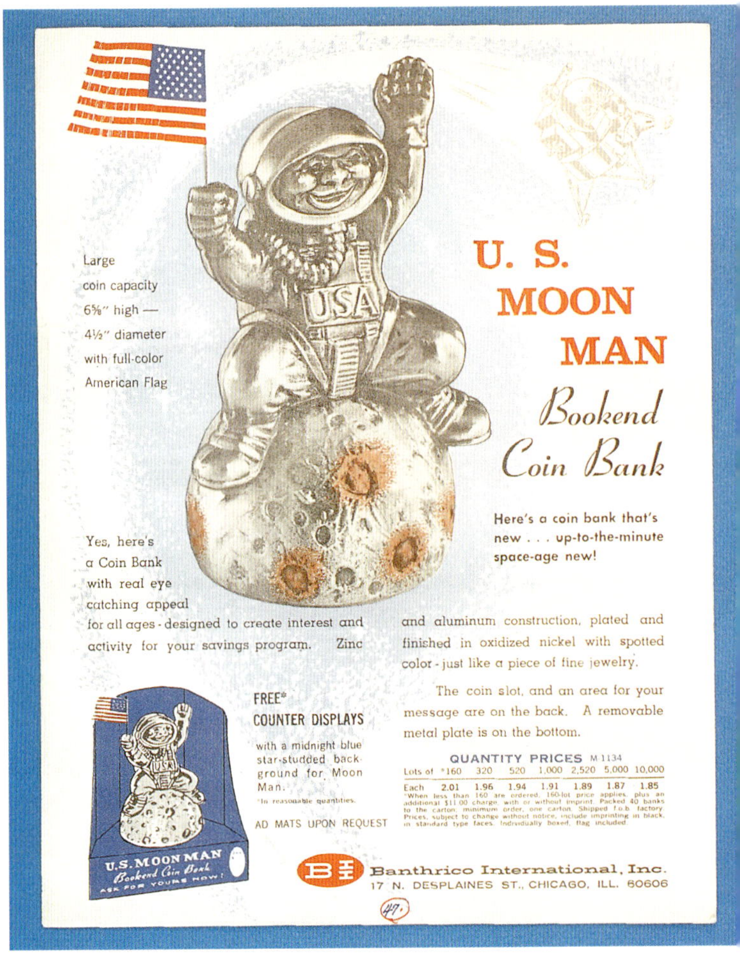

1021. Astronaut on Moon bookend bank

1022. American Heritage Collection banks

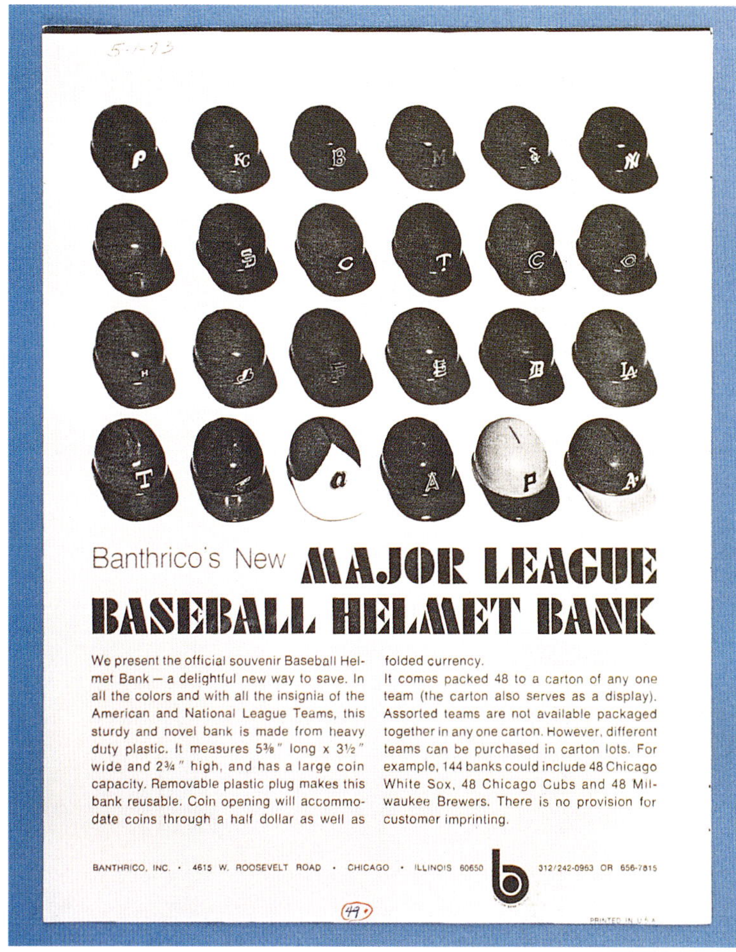

1023. Major league baseball helmet banks

1024. National Hockey League puck banks

1025. Hockey puck banks of different teams

1026. Plastic Mechanical "Jack in the Bank"

1027. "Little Lulu" vinyl bank

1028. Santa on roof with bag of toys

1029. Vinyl mechanical fictional character banks

1100. Steel buildings came in several color combinations

1101. Early lead and plastic banks

1102. 1940s-'50s popular white metal banks

1103. Early building, disk, and transportation banks

1104. Building, disk, and tower banks. Do you have the Wilson Building bank?

1105. White metal bank craftsmanship at its best

1106. More early banks with elusive owl bank

1107. The seated Indian boy is a great painted bank; note map of Idaho on log

1108. The bell is ringable and the Southwestern S & L building is an "E" bank

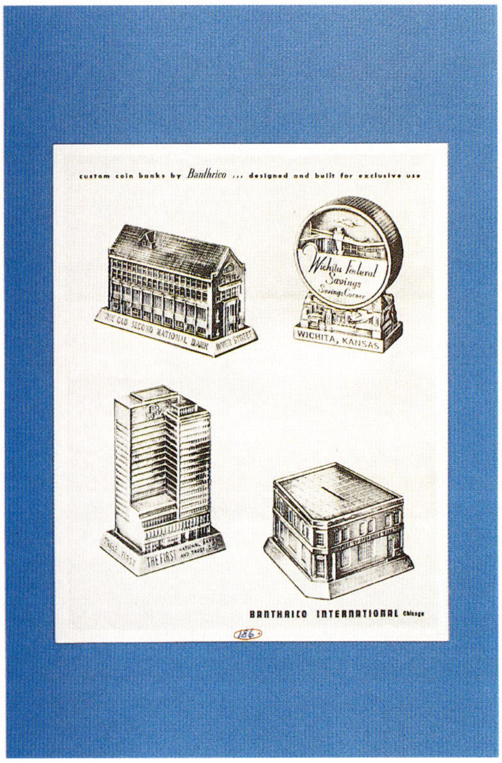

1109. Note airplanes on Wichita disk bank, which is rare for Banthrico

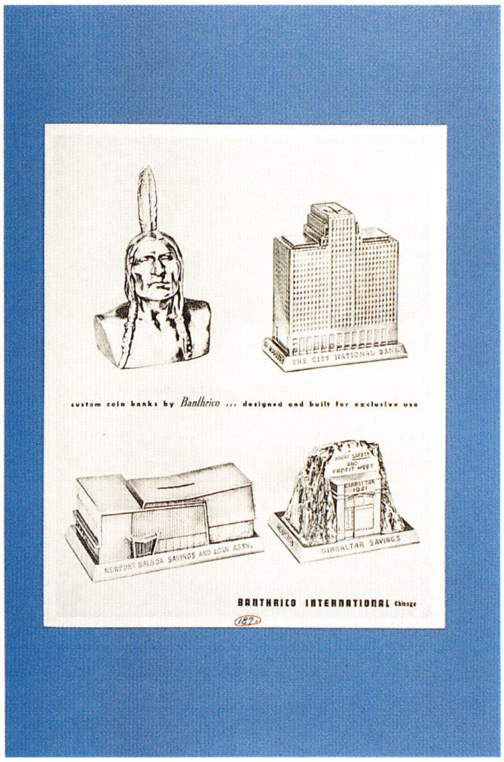

1110. Unusual Gibraltar Building like the "Rock of Gibraltar"

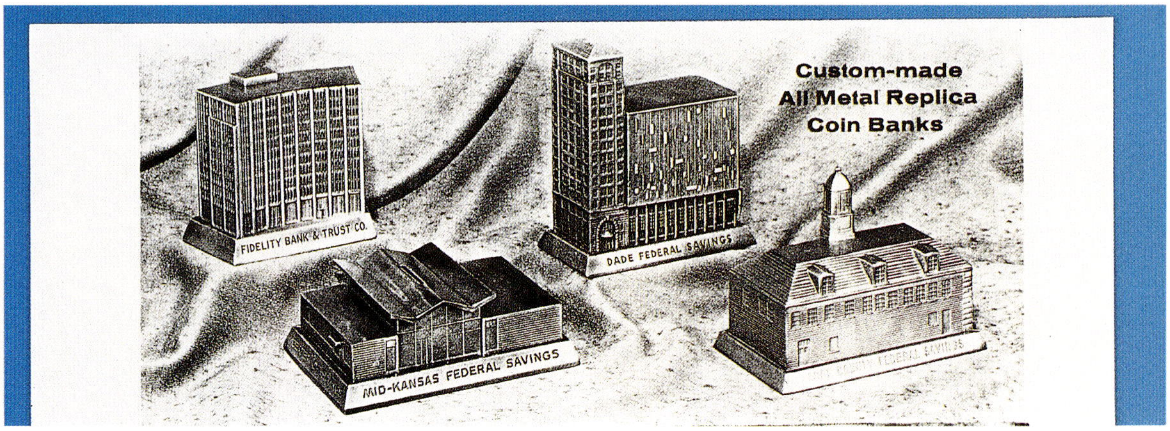

1111. Early Banthrico buildings

1112. Souvenir Building Collector's delight

1113. First Security bank, Salt Lake City, Utah

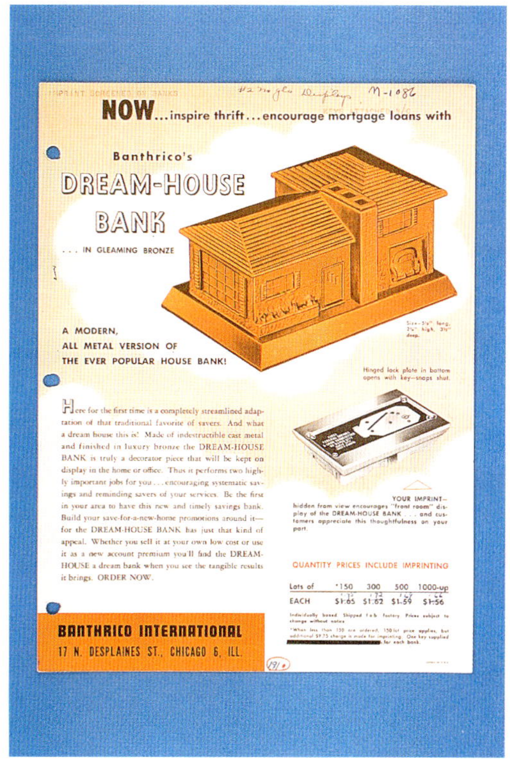

1114. Banthrico's dream house with garage and car

1115. More beautiful banks, paperweights, and letter openers

1116. The Oregon pioneer atop the Capitol building

1117. The Oregon pioneer arriving by truck

1118. Banthrico advertising displays

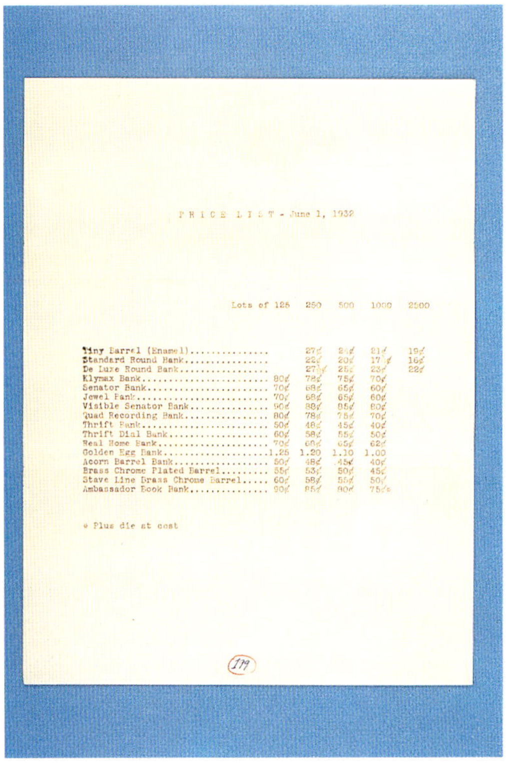

PRICE LIST - June 1, 1932

	Lots of 125	250	500	1000	2500
Tiny Barrel (Enamel)...............		27¢	24¢	21¢	19¢
Standard Round Bank...............		22¢	20¢	17½¢	16¢
De Luxe Round Bank...............		27½¢	25¢	23¢	22¢
Klymax Bank........................	80¢	78¢	75¢	70¢	
Senator Bank........................	70¢	68¢	65¢	60¢	
Jewel Bank...........................	70¢	68¢	65¢	60¢	
Visible Senator Bank..............	90¢	88¢	85¢	80¢	
Quad Recording Bank..............	80¢	78¢	75¢	70¢	
Thrift Bank..........................	50¢	48¢	45¢	40¢	
Thrift Dial Bank....................	60¢	58¢	55¢	50¢	
Real Home Bank....................	70¢	68¢	65¢	62¢	
Golden Egg Bank...................	1.25	1.20	1.10	1.00	
Acorn Barrel Bank.................	50¢	48¢	45¢	40¢	
Brass Chrome Plated Barrel.........	55¢	53¢	50¢	45¢	
Stave Line Brass Chrome Barrel.....	60¢	58¢	55¢	50¢	
Ambassador Book Bank..............	90¢	85¢	80¢	75¢*	

* Plus die at cost

1119. Banthrico 1932 price list

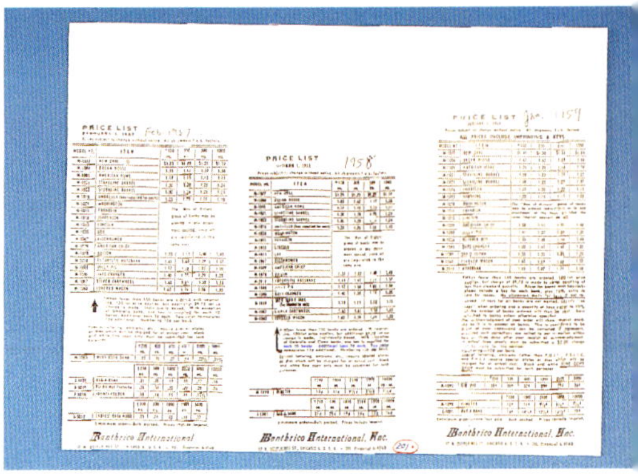

1120. 1957-'58-'59 Banthrico bank price lists

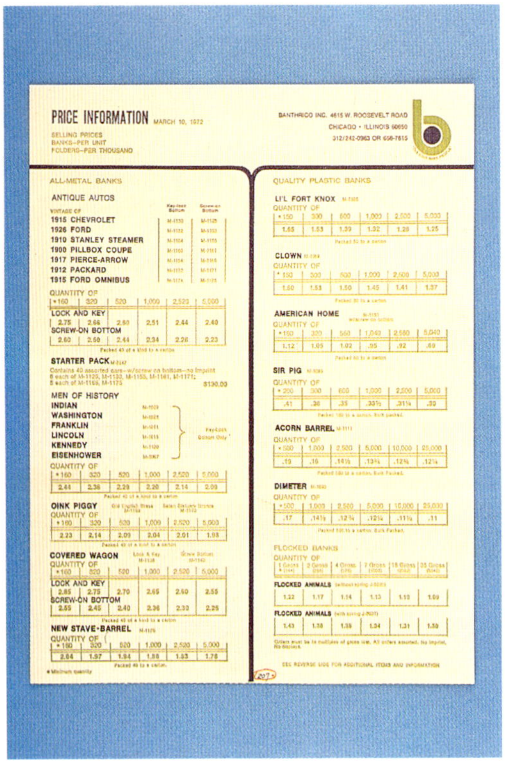

1121. 1972 price lists and traps available

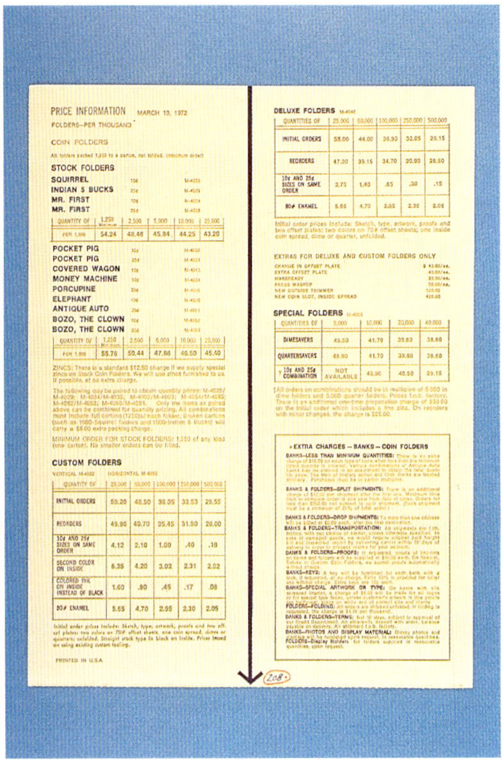

1122. Continuation of 1972 price information

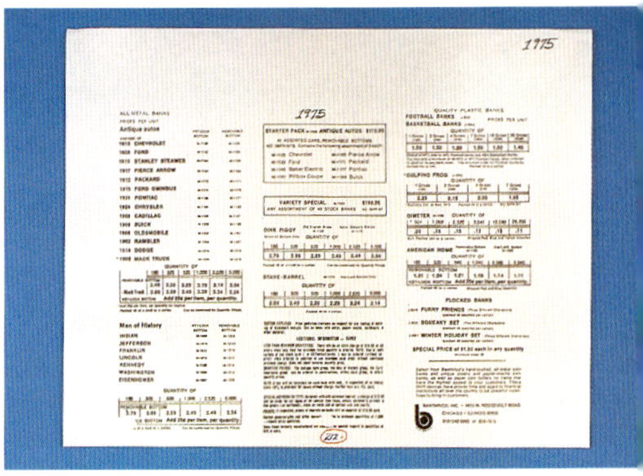

1123. 1975 price list and traps available

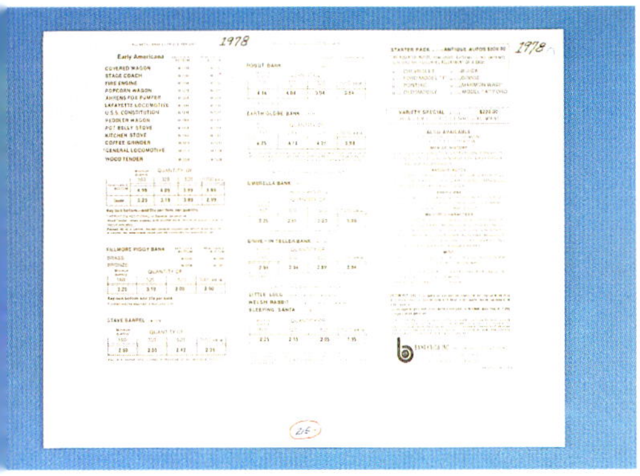

1124. 1978 price list. Starter packs. Note Kansas Herky mascot bank

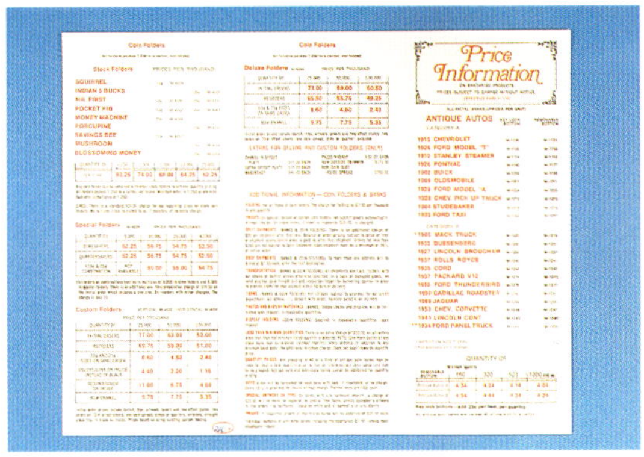

1125. 1980 price information and traps available

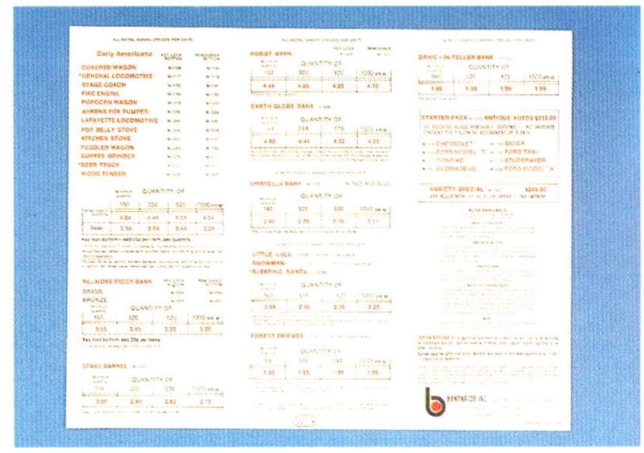

1126. Continuation of 1980 price information

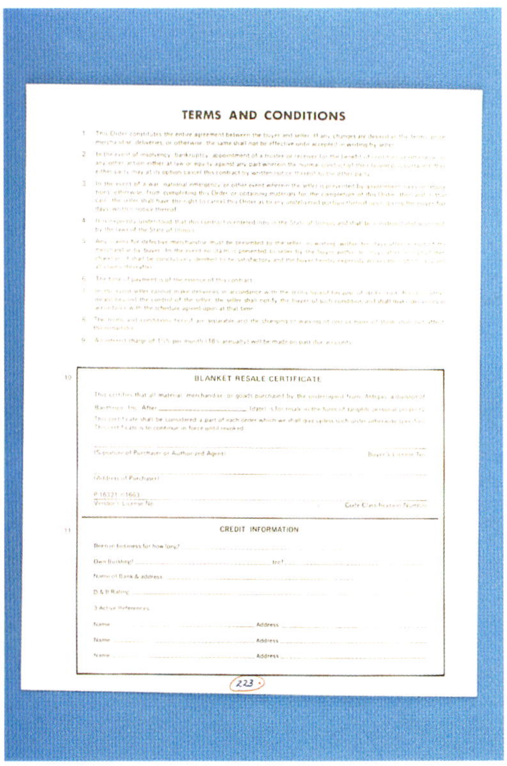

1127. Banthrico purchasing agreement

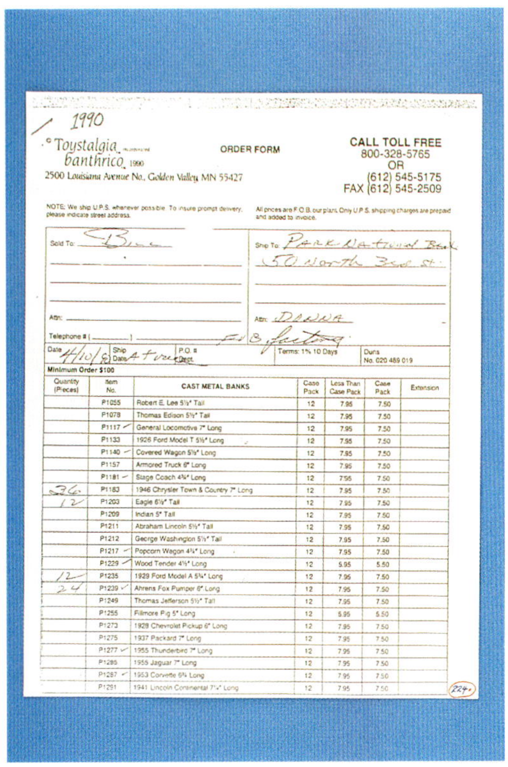

1128. 1990 Toystalgia order form and price list

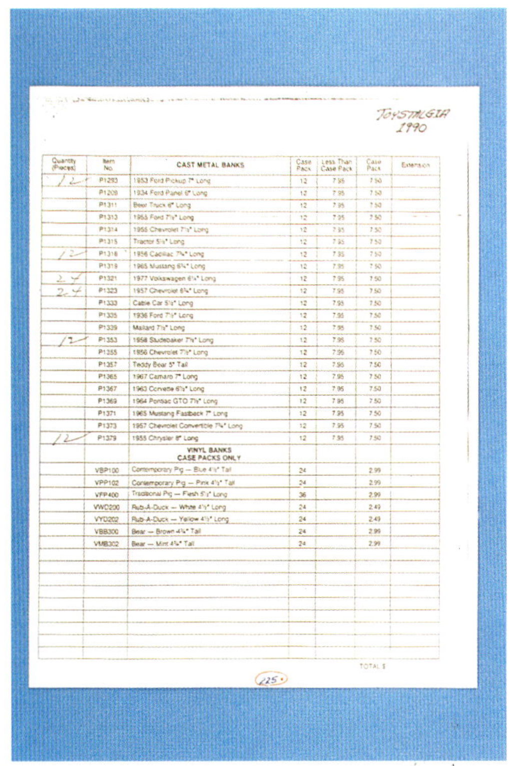

1129. Continuation of Toystalgia order form

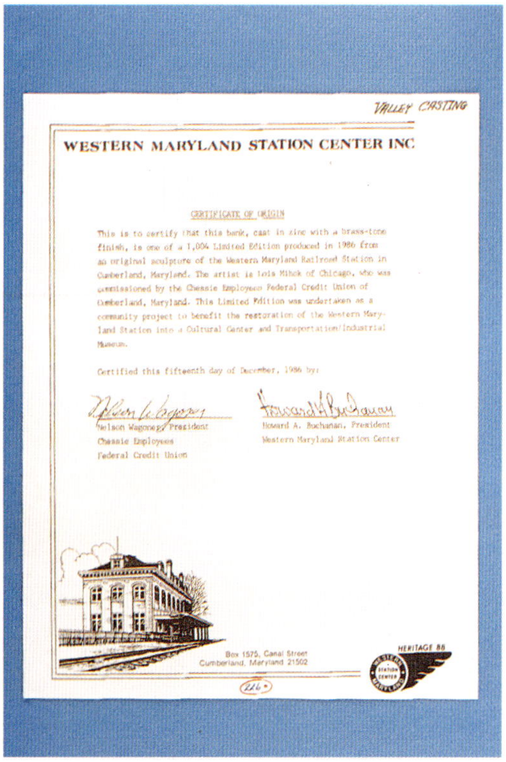

VALLEY CASTING

WESTERN MARYLAND STATION CENTER INC

CERTIFICATE OF ORIGIN

This is to certify that this bank, cast in zinc with a brass-tone finish, is one of a 1,004 Limited Edition produced in 1986 from an original sculpture of the Western Maryland Railroad Station in Cumberland, Maryland. The artist is Lois Mihok of Chicago, who was commissioned by the Chessie Employees Federal Credit Union of Cumberland, Maryland. This Limited Edition was undertaken as a community project to benefit the restoration of the Western Maryland Station into a Cultural Center and Transportation/Industrial Museum.

Certified this fifteenth day of December, 1986 by:

Nelson Wagoner, President
Chessie Employees
Federal Credit Union

Howard A. Buchanan, President
Western Maryland Station Center

Box 1575, Canal Street
Cumberland, Maryland 21502

HERITAGE 86

1130. Western Maryland station Center bank certificate

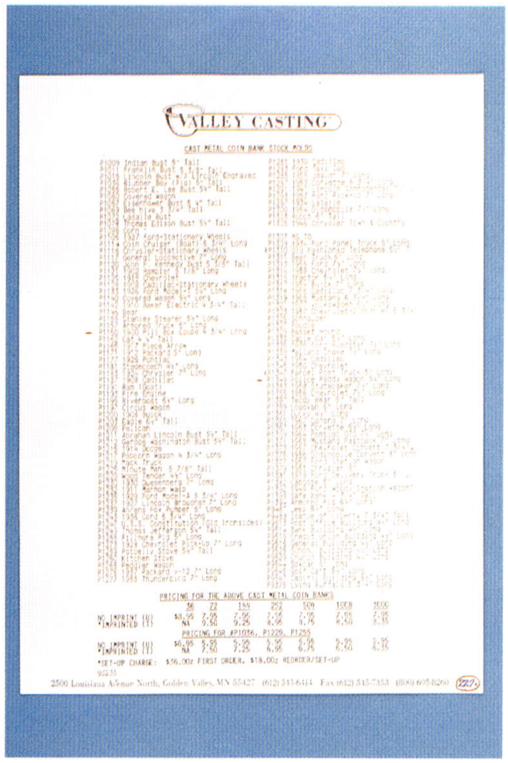

VALLEY CASTING

CAST METAL COIN BANK STOCK MOLDS

PRICING FOR THE ABOVE CAST METAL COIN BANKS

2500 Louisiana Avenue North, Golden Valley, MN 55427

1131. Valley Casting cast metal coin bank stock molds

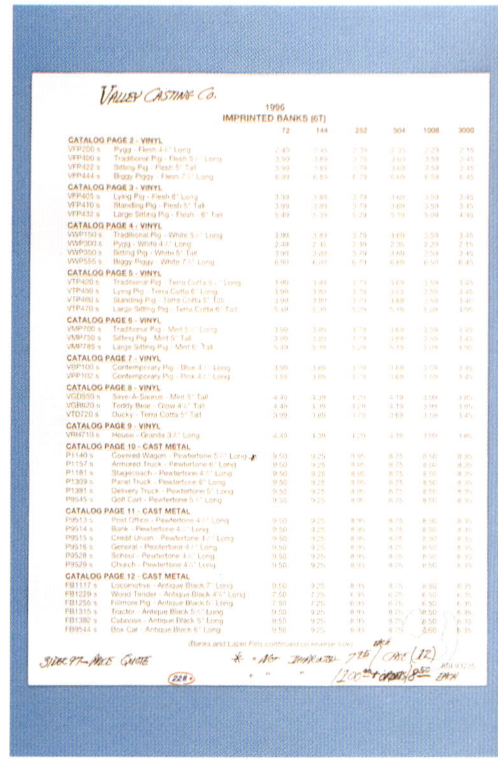

VALLEY CASTING CO.

1996
IMPRINTED BANKS (6T)

1132. 1996 Valley Casting imprinted banks list

1133. 1999 Valley Casting catalog page of custom banks and special runs. Can you identify the first bank of the second row?

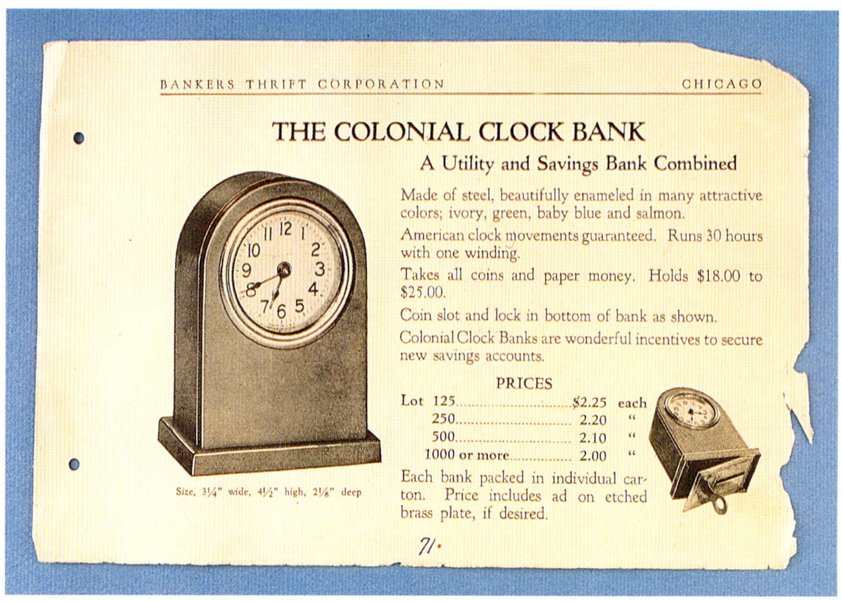

BANKERS THRIFT CORPORATION CHICAGO

THE COLONIAL CLOCK BANK

A Utility and Savings Bank Combined

Made of steel, beautifully enameled in many attractive colors; ivory, green, baby blue and salmon.

American clock movements guaranteed. Runs 30 hours with one winding.

Takes all coins and paper money. Holds $18.00 to $25.00.

Coin slot and lock in bottom of bank as shown.

Colonial Clock Banks are wonderful incentives to secure new savings accounts.

PRICES

Lot		
125	$2.25	each
250	2.20	"
500	2.10	"
1000 or more	2.00	"

Each bank packed in individual carton. Price includes ad on etched brass plate, if desired.

Size, 3¼" wide, 4½" high, 2⅛" deep

1134. Colonial Clock Bank

COIN CRUISER Bank (M-1114)

SIZE: 6¾" long 2⅞" high, 3" wide

Sleek, Sporty and Saveworthy!

This Coin Cruiser bank was designed for the smoothest thrift promotion ride yet. Here, indeed, is a decorative piece that will be kept on display and used in the home or office, by child or adult. This slick craft is constructed of all cast metal and finished in rich bronze plating. Be the first in your area to offer this new and timely savings bank.

LOTS OF	150*	300	500	1,000	2,500	5,000	10,000
EACH	$1.79	$1.75	$1.72	$1.69	$1.67	$1.65	$1.63

* A $9.75 charge is made for less-than-standard quantity. Shipped f.o.b. factory. Prices subject to change without notice. Prices include imprinting in white, in standard type faces, bank individually boxed, with key. Ad mats and counter displays in reasonable quantities on request. Special lettering or artwork charged for at actual cost.

COLOR: Bronze Plating

Imprint Area on rear of Boat for your silk-screened message.

Banthrico International, Inc. 17 N. DESPLAINES ST., CHICAGO 6, ILLINOIS

1135. Coin Cruiser Bank

Bibliography

Automobile Quarterly's General Motors: The First 75 Years. New York: Crown Publishers, Inc., 1983.

Bellows, Ina Hayward, *Old Mechanical Banks*. Chicago: Lightner Publishing Co., 1940.

Bender, Paul, "The Banthrico Story," Model Cars Magazine, 1970.

Bender, Paul A., *Model Car Promotional and Kit Guide*. Elkhart, IN: Brasilia Press, 1994.

Brown, Dort Fratzke, *Souvenir Buildings*. Sacramento, CA: Indisota Publishers, 1977.

Brown, Dort Fratzke, *Souvenir Buildings - Volume II*. 1979.

Butler, Steve, *Promotionals, 1934-1983*. Gas City, IN: L-W Book Sales, 1997.

"Coin Banks", Mid-Western Banker Magazine, 1982.

Cranmer, Don, *Collectors Encyclopedia of Toys and Banks*, Marion, IN: L-W Books, 1983.

Duer, Don, *A Penny Saved - Still and Mechanical Banks*. Atglen, PA: Schiffer Publishing Ltd., 1993.

Duer, Don, *Penny Banks Around the World*. Atglen, PA: Schiffer Publishing Ltd., 1997.

Egelhoff, Jeanie, Compiler/Historian, *Still Bank Collectors Club of America, Penny Bank Post Index, 1966-1997*. Indianapolis, IN: 1998.

Encyclopedia of Cars. New York: Mallard Press, 1990.

Gunnell, John A., Editor, *Standard Catalog of American Cars, 1946-1975*. Iola, WI: Krause Publication, 1982.

Jordan, Roy, *100 Banthrico Building & Mascot Banks*. Cedar Grove, TN: 1997.

Long, Ernest and Ida; Pitman, Jane, *Dictionary of Still Banks*. California: Long's Americana, 1980.

Majua, Margaret and Weingarten, David, *Souvenir Buildings - Miniature Monuments*. New York: Harry N. Abrams, Inc., Publishers, 1996.

Majua, Margaret, Trainer, Dixie and Hoffman, Barry, "Banthrico Building Banks." Nellysford, VA: Souvenir Building Collectors Society Journal, Fall 1996.

Majua, Margaret and Weingarten, David, *Monumental Miniatures*. Dubuque, IA: Antique Trader Books, 1999.

Moore, Andy and Susan, *The Penny Bank Book*. Atglen, PA: Schiffer Publishing Ltd., 1984.

Morgan, James, *Our Presidents*. New York: The MacMillan Company, 1959.

Oldham, Lance, "The Banthrico Story - Continued". Winter Park, FL: SBCCA Penny Bank Post Magazine, Summer 1997.

Oldham, Lance G., *Banthrico/Toystalgia/Valley Casting Co. Building/Monument/Landmark Bank Replicas*. Poulsbo, WA: 1998.

Peirce, Bob and Shirley, *Iron Safe Banks*. West Allis, WI: Still Bank Collectors Club of America, 1991.

Redwine, Jim, "Banthrico Banks are Hot!". New Berlin, WI: SBCCA Penny Bank Post Magazine, April 1998.

Rogers, Carole, *Penny Banks, A History and a Handbook*. New York: E.P. Dutton, 1977.

Webster's New Collegiate Dictionary, Colleges and Universities Section. Springfield, MA: G&C Merriam Co., 1979.

Whiting, Hubert B., *Old Iron Still Banks*. Vermont: Forward's Color Productions, Inc., 1968.

Closing Thoughts: 86,400 Seconds

Imagine there is a bank that credits your account each morning with $86,400. It carries over no balance from day to day. Every evening deletes whatever part of the balance you failed to use during that day. What would you do? Draw out every cent, of course!

Each of us has such a bank. Its name is TIME. Every morning, it credits you with 86,400 seconds. Every night it writes off, as lost, whatever of this you have failed to invest to good purpose. It carries no balance. It allows you no overdraft. Each day it opens a new account for you. Each night it burns the remains of the day. If you fail to use the day's deposits, the loss is yours. There is no going back. There is no drawing against the "tomorrow". You must live in the present on today's deposits. Invest it so as to get from it the utmost health, happiness, and success! The clock is running. Make the most of today.

To realize the value of One Year, ask a student who failed a grade. To realize the value of One Month, ask a mother who gave birth to a premature baby. To realize the value of One Week, ask the editor of a weekly newspaper. To realize the value of One Hour, ask two lovers who are waiting to meet. To realize the value of One Minute, ask a person who just missed the train. To realize the value of One Second, ask a person who just avoided an accident. To realize the value of One Millisecond, ask the person who won a silver medal in the Olympics.

Treasure every moment that you have! And treasure it more because you have shared it with someone special, special enough to spend your time. And remember that time waits for no one. Yesterday is history. Tomorrow is mystery. Today's a gift. That's why it's called the present!

Author Unknown

Index

The number(s) following the bank name are the bank's number(s) in this book.

D

E

F

V

W